POISONS

Also by Peter Macinnis

Bittersweet: The Story of Sugar

Rockets: Sulfur, Sputnik, and Scramjets

POISONS

*From Hemlock to Botox and
the Killer Bean of Calabar*

PETER MACINNIS

Arcade Publishing • New York

FIRST NORTH AMERICAN EDITION 2005

First published in Australia by Allen & Unwin under the title *The Killer Bean of Calabar and Other Stories*

Library of Congress Cataloging-in-Publication Data

    Macinnis, Peter.
        Poisons : from hemlock to botox to the killer bean of calabar / by Peter Macinnis.— 1st North American ed.
        p. cm.
    ISBN-10: 1-55970-761-5 (hc)
    ISBN-13: 978-1-55970-761-9 (hc)
    ISBN-10: 1-55970-810-7 (pb)
    ISBN-13: 978-1-55970-810-4 (pb)
    1. Poisons—Popular works. I. Title.

    RA1213.M33 2004
    615.9—dc22   2004029762

Published in the United States by Arcade Publishing, Inc., New York
Distributed by Hachette Book Group USA

Visit our Web site at www.arcadepub.com

10 9 8 7 6 5 4 3 2

EB

PRINTED IN THE UNITED STATES OF AMERICA

CONTENTS

DRAMATIS PERSONAE

Banks, Joseph (Sir) James Cook's botanist on the cruise of the *Endeavour* around the Pacific and up the east coast of Australia.

Bassawur Singh In the mid-nineteenth century, drugged his victims with datura seeds and robbed them, until he poisoned himself by accident and robbed no more.

Bocarmé, Count de Poisoned his brother-in-law with nicotine and was executed in 1849. His wife was acquitted.

Bonaparte, Napoleon Corsican-born French soldier, general, consul, emperor, and possible poison victim at his death in 1821.

Cadwalader, Thomas In 1745, identified lead contamination of rum as a cause of stomach gripes.

Claudius, Tiberius Claudius Nero Roman emperor who was poisoned by his wife, Agrippina, in AD 54. His son Britannicus was poisoned by Nero in AD 55.

Cleopatra Queen of Egypt who died, reputedly of self-inflicted snakebite, in 30 BC.

Crippen, Dr. Hawley Harvey Poisoned his wife with hyoscine and fled with his lover, Ethel Le Neve, but was captured thanks to the radio telegraph. He was hanged in 1910.

Cromwell, Oliver   Roundhead leader who refused to take quinine, not because it was poisonous but because Jesuits had brought it to Europe. He died of malaria in 1658.

Crookes, William (Sir)   Discovered thallium in 1861, and later used assorted poisons to fight the British cattle plague.

de la Pommerais, Couty   Poisoned his mistress with digitalis for a life insurance policy, and went to the guillotine in 1864.

Defoe, Daniel   Author of *Robinson Crusoe*, he described the use of various poisons in medicine.

Dioscorides, Pedanius   A botanist and knowledgeable herbalist who served in Nero's armies and died in AD 90.

Domitian, Titus Flavius Domitianus   Roman emperor who poisoned his niece while trying to abort their child.

Elisha   Old Testament prophet who knew a bit about preventing poisoning. He flourished around 850 BC.

Galba, Servius Sulpicius   Made emperor of Rome in AD 68 and assassinated in AD 69, executed Locusta and earlier distinguished himself by crucifying a Roman citizen who had murdered by poison.

Gnaeus Domitius   Great-great-great-grandfather of the later Roman emperor Nero. He tried to commit suicide by poison but reneged, in 80 BC.

Goldberger, Joseph   By rigorous testing, showed in 1914 that pellagra is a deficiency disease.

Harley, John   In 1867, tested hemlock on himself to establish its symptoms, and later recommended it as a treatment for hyperactive children.

Herodotus, of Halicarnassus   The first historian, who tended to repeat all that he heard. He died in about 425 BC.

Hippocrates, of Chios   The part-legendary traditional founder of medical science. He died around 377 BC.

Holmes, Oliver Wendell   American medical writer and physician who died in 1894; father of jurist of same name.

Humbug Billy William Hardaker, a Bradford candy seller whose wares had been accidentally filled with arsenious trioxide, killing 20. He wasn't even charged.

Leichhardt, Ludwig Australian explorer and adventurous taster who disappeared mysteriously in the 1840s. Did he taste the wrong thing, or did a flood get him?

Lister, Joseph English surgeon who first practiced antiseptic medicine, using poison against germs in about 1865.

Locusta A native of Gaul who sold poisons in Rome, and was eventually executed by Emperor Galba.

Louis XIV French king who was almost poisoned by his doctors as a young man and died in 1715 at the age of 77, having ruled over the most poison-ridden court in Europe.

Luce, Clare Boothe Wife of media magnate Henry Luce, writer, U.S. ambassador to Italy, and accidental poison victim, thanks to flaking arsenical paint.

Marsh, James Infuriated by the acquittal of a guilty party, developed the exquisitely sensitive Marsh test for arsenic in corpses.

Maybrick, Florence Convicted in 1889 of murdering her husband, James, with arsenic, she was released in 1904.

Mithridates King of Pontus who, according to legend, took poison in small doses to make himself immune to poisons. He died in 63 BC.

Moore, J(oseph) Earle Medical practitioner and writer on syphilis, who died in 1957.

Muro y Fernandez-Cavada, Dr. Showed that a Spanish poisoning case, attributed to contaminated oil, may have had another cause.

Ovid, Publius Ovidius Naso Roman poet who was a bit of a gossip, and who at one stage was banished to Pontus before dying in AD 17.

Palmer, William "The Rugely murderer," who may have killed several people or none with strychnine. He was hanged in 1854.

Pliny (the Elder), Gaius Plinius Secundus Roman writer, scholar, and fleet commander who was asphyxiated by poison gases from Mount Vesuvius in AD 79.

Portinari, Candido Brazilian painter who died in 1962, poisoned by his pigments.

Pritchard, Edward Poisoned his wife, and probably his mother-in-law, with antimony and was hanged in 1865.

Roose, Richard Attempted to kill a bishop with arsenical porridge and was boiled to death in 1531.

Scheele, Carl Discovered prussic acid, and was killed by it four years later in 1786.

Seddon, Frederick Poisoned his lodger, Eliza Barrow, after selling her an annuity. He was hanged in 1912.

Sherman, Mary Acquitted at the Old Bailey in London in 1726 of an arsenic murder. She may have been saved by the primitive tests available.

Shipman, Harold Medical poisoner who committed suicide in 2004, and who may have killed over 200 elderly patients with lethal doses of morphine.

Smith, Madeleine Acquitted on a "Not Proven" verdict of the murder of her lover by arsenic in 1857.

Socrates Greek philosopher who was condemned to death by taking hemlock in 399 BC.

Suetonius, Gaius Suetonius Tranquillus Roman historian who told a lot of how it really was in the courts of the Roman emperors. He died in AD 160.

Tawell, John A convicted forger who was transported to Australia. Tawell returned to England and later poisoned his mistress with cyanide. He was hanged in 1845.

Wainewright, Thomas A forger who was suspected of poisoning several relatives who died conveniently in the 1820s.

Xenophon Greek soldier, writer, historian, died 354 BC.

Yushchenko, Viktor Ukrainian politician, almost certainly poisoned with dioxin by rivals in 2004, survived.

GLOSSARY OF POISONS

Aconite   Two plants share this name: one is a daisy, the other is a member of the buttercup family. The name is also used to refer to the alkaloid derived from the root of the buttercup relative, which is also called monkshood or wolfsbane.

Aconitine   A drug and also a poison obtained from aconite.

Agaric   A comparatively harmless tree fungus, and fly agaric, a deadly mushroom.

Alcohol   A poison even in small doses, but one that luckily makes most of us ill before we can swallow a deadly dose. Hardened drinkers can fight this impulse and die as a result.

Alkaloid   A nitrogen-containing base produced by plants. Most alkaloids are biologically active, with many of them being powerful poisons, just what herbivores don't want. Morphine, codeine, nicotine, cocaine, hyoscyamine, ephedrine, strychnine, and atropine are all alkaloids.

Aniline   Otherwise known as aminobenzene, this is a toxic industrial solvent that can be absorbed through the skin, in food, or by inhalation.

Anticoagulants   These are not really poisons in the strict sense but kill by stopping the blood from clotting. They are used against rats and mice by poultry producers, because birds are hardly affected. They are also used in some medical treatments.

Antimony A heavy metal that is toxic both as a metal and as many of its compounds.

Arsenic A heavy metal poison, usually given as the oxide. It used to be common in cosmetic preparations, and some people even ate it "to improve the wind." It was easy to obtain in the nineteenth century, and easy to detect with the Marsh test.

Atropine See **Belladonna**.

Belladonna The deadly nightshade, belladonna, produces atropine, otherwise known as daturine. This speeds the heart rate, and causes delusions and delirium.

Bitter apple See **Colocynth**.

Botox A toxin produced by an anaerobic bacterium, *Clostridium botulinum*, also used for medical and cosmetic purposes.

Cacodyl cyanide Dimethyl arsenic cyanide, which gives off a poisonous vapor when exposed to air. An explosion of this compound took out one of Robert Bunsen's eyes. Its use was proposed in the Crimean War, but dismissed as too uncivilized.

Calabar bean See **Ordeal bean**.

Cantharides or **Spanish fly** A poison reputed to provoke overwhelming lust but better established as a dangerous toxin. It is prepared by soaking prebruised, freshly macerated beetles in chloroform. What were those people thinking about?

Carbon dioxide More of a smothering agent than a poison; this gas has been known to kill by depriving victims of oxygen.

Carbon monoxide   A gas produced by inefficient burning of carbon-based fuels, which locks on to hemoglobin in the blood and stops it from carrying oxygen or carbon dioxide.

Cardiac glycosides   Steroid compounds that do most damage to the heart and the kidney. They are found in a number of plants, and seem mainly to have a herbivore-repelling role.

Chlorine   A poison gas used in World War I, it was corrosive and toxic and choked its victims.

Chrome yellow   Otherwise known as lead chromate, this is toxic, though not greatly so in the doses used to color food.

Colocynth or **Bitter apple**   Known as far back as biblical times, it was used as a fairly drastic medicine, but it could also kill.

Cyanide   Hydrogen cyanide or prussic acid is much loved by the writers of murder mysteries and a deadly poison because it stops the blood passing oxygen to the cells. Sodium and potassium cyanide are also dangerous poisons.

Cycads   Primitive, cone-bearing palm-like plants with poisonous seeds.

Datura   Properly *Datura stramonium*, also called Jimsonweed or Devil's Trumpet, this plant produces atropine and scopolamine, with the seeds, fruit, and leaves having the highest concentrations of these alkaloids.

DDT   An insecticide that is also harmful to some wildlife, but no serious effects on humans have yet been discovered.

Diethylene glycol   A common component of antifreeze, once thought to be harmless, but this solvent is broken down by alcohol dehydrogenase to make poisonous oxalic acid. The cure is to treat the patient with alcohol!

Digitalis   The common foxglove produces a number of toxins with similar names: digitalin, digitalein, digitonin, and digitoxin. All are poisonous, though some have medical uses.

Dioxins   A family of organic compounds containing chlorine. They appear to damage DNA in some way and certainly seem to affect the offspring of those exposed. Dioxins are also formed as impurities in combustion such as peat fires.

Ergot   A fungal infestation in grasses that may produce as many as 20 different toxins.

Heavy metal   A member of a group of elements with similar chemistry, including lead, arsenic, antimony, mercury, and cadmium. All of them are toxic, both as metals and even more as compounds, and all of them tend to be accumulated in tissues and also up the food chain.

Hemlock   A parsley-like plant that produces a gentle death, quite unlike the violent death caused by the unrelated "water hemlock." It is most famous as the state poison of the ancient Greeks, and the cause of death of Socrates.

Laudanum   A tincture of opium, popular in the nineteenth century as a medicine and as a recreational drug, and occasionally used as a poison.

Lead   A toxic heavy metal, as are many of its salts. Lead breaks the disulfide bridges in proteins, which changes their shape and blocks their action.

Levant nut   *Cocculus indicus* was used by thieves and assassins to knock out their victims. It was also used by unscrupulous publicans to produce a stupefying effect in watered beer.

Mercury   A heavy metal poison mostly encountered in industrial situations. It can also be accumulated in various fish to dangerous levels.

Methyl isocyanate   An intermediate in the production of insecticides; the disaster at Bhopal established it as both toxic and corrosive.

Mushrooms   Many species are poisonous, some more so in the presence of alcohol. Fly agaric is used both as an intoxicant and to kill flies.

Mustard gas   2,2-dichloroethyl sulfide was used as a poison gas in World War I. It settled on materials and poisoned by contact and was a useful way of making certain areas off-limits.

Nerve gases   Assorted poisons that act by interfering with the transmission of nerve impulses, paralyzing the victim.

Nerve poisons   See **Organophosphates**.

Nicotine   Harmful enough in cigarette smoke, this alkaloid can be deadly when swallowed or absorbed through the skin.

Ordeal bean   The bean of *Physostigma venenosum*, a member of the pea family that contains a powerful poison, physostigmine. The lethal dose seems to be around a quarter of a bean, but in West Africa a test of truth-telling was to eat half a bean and survive.

Organophosphates   A group of common insecticides that work by interfering with the transmission of nerve impulses.

Oxygen   This gas is poisonous to anaerobic bacteria but so essential to us that many poisons work by impeding our access to it.

PCBs   Polychlorinated biphenyls, once thought to be harmless enough, but not anymore. They accumulate in the food chain and appear to be fetotoxic.

Penicillin   A deadly poison to bacteria but less so to us.

Phosgene   This compound of chlorine and carbon monoxide was used as a poison gas in World War I. It caused a buildup of fluid in the lungs, and victims drowned.

Phosphorus   White phosphorus is highly poisonous and attacks many of the body's organs. It was used as a rat poison but seems to have been little used against humans.

Prussic acid   See **Cyanide**.

Quinine   A poison in large doses, in smaller doses it kills malarial parasites.

Ricin   A poison produced in the castor bean. A single molecule is enough to knock out the ribosomes in a cell, killing the cell over time. There is a lethal dose from as little as three seeds.

"Roger"   A pet name for chlorine coined by English soda industry workers.

Rotenone   A garden pesticide obtained from the ground root of the Jamaica dogwood. It was once used to stupefy fish to make them easier to catch.

Sodium fluoroacetate   Also known as "1080," this poison is used in baiting to kill rabbits and foxes.

Strychnine   An alkaloid obtained from the dried ripe seeds of *Strychnos nux-vomica*. It was a common rat poison and also a tonic! It can build up in the system and is reputed to have killed the famous racehorse Phar Lap.

Tetrodotoxin   Also known as TTX, this toxin is common in many life forms. It is probably made by an as yet unknown single-celled organism, possibly a bacterium, and is passed along in the food chain.

Thallium   Another heavy metal, which has the curious side effect of making the victim's hair fall out, hence its occasional use as a depilatory. Its toxic properties made it a common rat poison, because it acted slowly, lulling rats into a sense of security, so they ingested a lethal dose.

Wormwood   This plant produces artemisinin, which can be used to give a "kick" to absinthe, to kill worms, and maybe even to kill malarial parasites.

PROLOGUE

Books often have peculiar provenances. I began this one in whimsy, chatting to Emma the Excellent Editor about Mr. Pugh, a character in Dylan Thomas's *Under Milk Wood*. Poor Mr. Pugh was a schoolteacher, and Thomas gives him a "moustache worn thick and long in memory of Doctor Crippen." Henpecked Pugh sat at the breakfast table, reading his book, *Lives of the Great Poisoners*, and dreaming of poisoning Mrs. Pugh.

I said in the middle of a conversation about something completely different that I was sure nobody had ever written a book called *Lives of the Great Poisoners*, and within a few minutes I was arguing to write *Mr. Pugh's Breakfast Table Book*. Later I saw that most of Mr. Pugh's Great Poisoners were abject failures, because they were found out. The clever ones, as Balzac pointed out, got away with it, eluding both punishment and fame.

> *Le secret des grandes fortunes sans cause apparente est un crime oublié, parce qu'il a été proprement fait.* (The secret of great fortunes for which you are at a loss to account is a crime that has never been found out, because it was properly executed.)
>
> Honoré de Balzac, *Le Père Goriot*, chapter 2

It mattered little. By then I was looking more widely at poisons, where they are found, why they do harm, and how they are used to do good. That led me to the evolution of poisons, medical poisons, the poison battle between toxic bacteria and drugs, and the battle played out in the nineteenth century between the poisoners and those seeking to stop poisoners, if only by demonstrating that any poison could be detected in a corpse, no matter how cleverly the poison was administered, and thus taking away the hope of eluding detection.

You can take almost any starting point and trace a trail of poison. Let me demonstrate. (All the instances mentioned here will be met later in this tale.) George Bernard Shaw was a good socialist who kept company with socialist writers and men of letters like H. G. Wells, Clive Bell, and Leonard Woolf: suppose we take this small and select group of Fabians as the epicenter and see how they were all associated with poisons in different ways.

Shaw met Madeleine Wardle, formerly the famous or infamous Madeleine Smith, who had been cleared of the arsenic murder of her lover by a "Not Proven" verdict. She had then married the Pre-Raphaelite painter George Wardle. Shaw said later that he found her quite pleasant. Shaw's friend Wells exposed the evils of lead poisoning in the potteries of England, where plumbism was rampant, while Leonard Woolf and Clive Bell were married to the Stephen girls, Virginia and Vanessa.

Virginia Woolf and Vanessa Bell had an uncle who was the judge in Florence Maybrick's trial for the murder by poison of her husband. Some people say that James Maybrick may have been Jack the Ripper; another unlikely suspect was the judge's son, the cousin of Virginia and Vanessa, the poet J. K. Stephen. Yet another Ripper suspect was the doctor and convicted poison murderer Neill Cream. Cream was a medical student with Arthur Conan Doyle, himself recently accused of committing a poison murder, almost a century after the death in question.

Florence Maybrick's husband and Madeleine Smith's lover were both reported to be addicted to arsenic, and another Pre-Raphaelite, Dante Gabriel Rossetti, married his model, Elizabeth Siddal, who used arsenic to make her complexion paler. She later poisoned herself with laudanum.

On top of this, George Wardle worked both for and with William Morris. Morris inherited a fortune that came from arsenic mining, a fortune that he added to by designing and selling wallpapers in which the main pigment was a salt of arsenic.

The poison chain can go on and on—we might mention yet another alleged candidate for Jack the Ripper, Lewis Carroll, who wrote (disapprovingly) of children drinking poisons in *Alice in Wonderland*: "if you drink much from a bottle marked 'Poison,' it is almost certain to disagree with you, sooner or later."

People find a delicious fascination in viewing at a safe distance the acts of psychopathic murderers and terrorists, but the most effective wielders of poisons around humans are bacteria and those who can manipulate them. They, in truth, are the best of the Great Poisoners, but we cannot understand them without understanding the more traditional poisons as well.

And so this book was born. No more is it Mr. Pugh's breakfast table book, an account of the lives of the great poisoners. It is a tiptoe among murderous herbs and minds, a chance to taste-test in complete safety some of the more interesting poisoners—and their poisons.

POISONS

1

POISON'S CHILDREN

Poisoners do their deadly work in secret. The evidence in poisoning cases is nearly always circumstantial. In this case it is more direct than usual. You must remember that poisoning is always concealed and deliberate. It is a crime that is not done in a moment of passion, or on an impulse. It is a crime that must be planned.

Justice William Windeyer, Central Criminal Court, Sydney,
summing up in the Dean case, April 6, 1895

On one level, poisoners are takers of lives; on another level, poisons cleverly applied should be the key to the major breakthroughs of medicine in the twenty-first century, just as they were in the twentieth. The poisoners of today generally act for the benefit of humanity, but we tend to ignore them as we concentrate on a small minority, the poison murderers.

Poisoning, done with deliberation against a fellow human, is planned, secret, and dramatic, and this makes it memorable. We recall the Cleopatra of legend who died with an asp clasped to her breast; we forget the thousands who had to die in battle to show her that it was time to bring on the serpent. We recall the

Borgias, who poisoned a few foes, and forget the many others of their era who relied on stiletto-, rapier-, or cudgel-wielding thugs to kill as many or more.

We recall and celebrate Mithridates, who made himself proof against poisoning by taking increasing doses, but what do we know of or care about those who sought to protect themselves by wearing armor? We remember Socrates, who was put to death by being made to drink hemlock, even as we forget many others who met the headsman, guillotine, garrotte, firing squad, or hangman.

Let us not forget that, in a world where men are generally larger and more belligerent than women and better able to handle weapons, poison was a weapon women could use just as well, and a weapon the weak could use against the strong. But best of all, poison left no bleeding, gaping holes if you selected the dose thoughtfully. In his plays, Shakespeare's characters had poison in their ears, their food, a chalice, a potion, or on the blade of a weapon. Whatever the method, the big plus was and is the way poison levels the playing field; moreover, used carefully, poison offers the poisoner every chance of escaping retribution. This makes poison something for men to fear; and psychologists say the things we fear are often found in the most popular children's tales, like the poisoned apple of Snow White and the evil drug of Captain Hook, who sets out to poison Peter Pan:

> Lest he should be taken alive, Hook always carried about his person a dreadful drug, blended by himself of all the death-dealing rings that had come into his possession. These he had boiled down into a yellow liquid quite unknown to science, which was probably the most virulent poison in existence.
>
> Five drops of this he now added to Peter's cup.
>
> James Barrie, *Peter Pan*, 1911

Then there are the folk tales, now on the Internet but which have been circulating for many years, of poison rings used by white slavers to drug young girls, envenomed needles dripping HIV, ATM envelopes and stamps treated with deadly poisons, and poisoned candies dropped by German zeppelins over Britain and France in World War I. Poison is everywhere in our imagination; it is also everywhere in the real world. We are generally unaware of the real poisons that surround us, and, given our reactions to the fictional ones, that is probably just as well.

Alice considers the bottle marked "Drink me."

Poisons are everywhere, if you look hard enough, and we can now blame evolution for this. If you are an animal, and something with pointy teeth and bad breath comes slurping and salivating in your direction, you run away. If you are a plant and some lumbering beast starts to bite chunks out of you, your best defense is to poison it.

Evolution shapes poisons in curious ways. For example, there is the problem of seed dispersal, where plants need animals to eat their fruits, pass the seeds through, and drop them somewhere else, in a small damp patch of dung—instant mulch, moisture, and nutrition for the sprouting seeds. Animals with grinding teeth may crush the seeds, so there is a delicate balance between encouraging animals to eat and sending them on their way, as we can see in the *Capsicum* family.

Capsaicin is unpleasant stuff: police use it to subdue the unruly, in the form of pepper sprays, but it turns out that capsaicin repels or poisons mammals but not birds. In a 2001 study, researchers found that peppers in the field were unpalatable to cactus mice and pack rats, but the curve-billed thrasher devoured them with impunity. The bird was the most effective at spreading the seeds without harming them, the researchers found.

Marijuana, *Cannabis indica*, makes its active ingredient, tetra-hydrocannabinol (THC), to discourage cattle from eating it, because they, unlike humans, do not like being stoned, though the Scythians did, according to Herodotus:

> On a framework of three sticks, meeting at the top, they stretch pieces of woolen cloth, taking care to get the joins as perfect as they can, and inside this little tent they put a dish with red-hot stones in it. Then they take some hemp seed, creep into the tent, and throw the seed onto the hot stones. At once it begins to smoke, giving off a vapor unsurpassed by any vapor-bath one could find in Greece. The Scythians enjoy it so much that they howl with pleasure. This is their substitute for an ordinary bath in water, which they never use.
>
> Herodotus, *Histories, c.* 430 BC

But if cattle fear intoxication, some animals actually depend on getting high. Australian koalas need very little water, because they take most of it from their food. They sit in Australian gum trees, looking cuddly, but in reality they are just spaced out on eucalyptus oil, out of their tiny little minds on the toxins lurking in the leaves of the gum tree—and their minds have to be tiny. Brains are major energy users and koalas simply cannot afford to use lots of energy, because that would mean eating more leaves, which would poison them even more.

Move to a new place, and a whole new range of poisons is available to you; safety lies in having a bit of local knowledge. This was known even in biblical times, when Elisha visited Gilgal:

> . . . and there was a dearth in the land; and the sons of the prophets were sitting before him: and he said unto his servant, Set on the great pot, and seethe pottage for the sons of the prophets.
>
> And one went out into the field to gather herbs, and found a wild vine, and gathered thereof wild gourds his lap full, and came and shred them into the pot of pottage: for they knew them not.
>
> So they poured out for the men to eat. And it came to pass, as they were eating of the pottage, that they cried out, and said, O thou man of God, there is death in the pot. And they could not eat thereof.
>
> But he said, then bring meal. And he cast it into the pot; and he said, Pour out for the people, that they may eat. And there was no harm in the pot.

It is widely agreed that the poison in this case was in the fruits of a local vine, highly regarded in small doses for its medicinal properties, and called either colocynth, bitter apple, or *Citrullus colocynthis* (L.) Schrad., if you happen to be a botanist. As long as you know how to deal with it, however, it does not matter much what you call it.

Human nature being what it is, as soon as the harmful properties of an object—be it animal, vegetable, or mineral—were established, people found ways to use it for nefarious purposes. Xenophon, for example, comments on Persia that

> Children in former times were taught the properties of plants in order to use the wholesome and avoid the harmful; but now they seem to learn it for the mere sake of doing harm: at any rate, there is no country where deaths from poison are so common.
>
> Xenophon, *Cyropedia*, 360 BC

Citrullus colocynthis

One of the more attention-getting poisons over the years has been arsenic, and we get the word from the Greek "arsenikon," which was the name the Greeks gave to orpiment, a yellow arsenic sulfide pigment, but the word stretches back into Syriac, Middle Persian, and Old Iranian. Everybody, it seems, liked orpiment, in part because its gold color reminded hopeful alchemists and others of gold. Indeed, the word *orpiment* is itself a corruption of the Latin *auripigmentum*, or gold pigment, reflecting its main non-medical use in decoration and art.

Livy records the first Roman judicial trial for poisoning as far back as 329 BC, and Suetonius, the Roman historian who told it like it was, claims one Gnaeus Domitius, the great-great-great-grandfather of Nero, tried in 80 BC to commit suicide by taking poison. Having second thoughts, he quickly demanded an emetic and vomited the poison up. The physician who supplied the poison was one of his slaves, and knew his master well enough to have given him only a mild dose, so Gnaeus Domitius granted him his freedom as a reward.

At that stage, the Romans had hardly begun to poison. By Ovid's time (he died in AD 17), everybody was hard at it, and the poet wrote of how men lived on plunder, much of it won by poisoning. Guest was not safe from host, he said, nor father-in-law from son-in-law. Even among brothers it was rare to find affection, while husbands longed for the deaths of their wives, who reciprocated, and murderous stepmothers brewed deadly concoctions,

Locusta was a lady of Gaul who gravitated to Rome, where she plied her trade as purveyor of fine poisons to the selected nobility. Occasionally her trade extended to consultation, and, for the squeamish and inexperienced, Locusta would even administer her poisons to selected targets, including some in the imperial family.

Suetonius names her as the source of the arsenic Nero used on Britannicus, the son of Claudius. Locusta provided too mild a dose, and Nero flogged her. Locusta protested that the weak dose was to make the death less obvious, but Nero, saying he was not frightened of the Julian law against poisoning, demanded a stronger dose, which he tried first on a kid, but it was still too slow.

After further boiling down, he tried it on a pig. When the pig dropped dead on the spot, he gave the rest to Britannicus the same night and rewarded Locusta with her freedom, though she had already been condemned to death as a poisoner. He later sent students to her. In the end, Locusta was killed by Galba when he became emperor.

and "sons inquired into their fathers' years before the time."

Deliberate poisoners aside, the amount of lead the Romans ingested on a daily basis would have made them poisoned and venomous. The lead was in the water and their wine, along with many natural poisons. Few of these natural poisons, however, tasted as sweet as lead acetate, often called sugar of lead, or *sapa* by the Romans, who added it to wine to improve the flavor—but it certainly made no improvement on them or their dispositions.

Lead acetate is uncommon in its sweetness, and one of the most obvious questions must be why so many poisons have a bitter taste. The answer probably has to do with the way so many venomous animals are brightly colored, to warn their predators off. "Don't start anything, don't tread on me," they say, and for the most part it saves them from having to defend themselves. Plants, on the other hand, can afford to lose a bit of their mass, and a taste with the same "Don't start anything" message is as good as a color warning, but a bad-tasting poison sends an even blunter message. A bad taste works against herbivores of all sizes and visual acuities, and faster than any poison, but poison provides permanent relief from those who won't take a hint.

This association of poison with a bitter taste is probably lucky for those of us who might otherwise fall prey to the wiles of the poisoners, who are driven to rely instead on the smaller number of almost tasteless poisons, often of a mineral origin. Even here, there is a sort of evolutionary process, with poisoners like Dr. Lamson (whom we will meet later) seeking out the undetectable poisons, and the good guys working as fast as they can to close off any loophole by finding ways to detect the subtle new poisons.

We evolved in poisonous surroundings, all of us—the humans, the beasts of the field, the birds, the fish, even the plants. Life probably evolved first at something like a hydrothermal vent, deep in the ocean. Scientists already know of 3.2-billion-year-old fossils at former vent sites, and unfriendly as it seems, a dark place with scalding water, noxious metals, and savage gases is probably the cradle of all life as we know it today. That makes us poison's children indeed.

We are also the descendants of a later gang of active poisoners, whose venom wiped a planet almost clean of all previous life forms. These mass murderers were aerobic life forms that evolved and started spewing a poison into the planet's atmosphere. This was oxygen, the gas produced when photosynthesis splits water as part of a process that hitches a free ride on sunlight, using it for food.

The brutally toxic oxygen was sopped up at first by other elements on the planet's surface, producing huge deposits of iron oxide, but, in the end, those sinks for the oxygen were all used up and still the oxygen kept coming, poisoning most of the anaerobic life forms. A few of them clung on and are still there, lurking in odd places and causing gangrene and tetanus when the conditions are right to let them get their own back. Others, like *Bacteroides gingivalis*, live in obscure places such as the tiny, oxygen-poor gaps between our teeth, until they are rudely thrust

There are other anaerobic bacteria, like the gas gangrene bacterium, *Clostridium perfringens*, which produces toxins to kill flesh to make more living space for itself, and the related *Clostridium botulinum*, a major cause of lethal food poisoning in foods like canned meat and sausages (*botulus* being the Latin for "small sausage"). Botulinum toxin, or Botox to the fashion-conscious, is generally rated as the most powerful poison around, and we will meet up with it again later. Whichever way you look at it, we need not have too much sympathy for the anaerobes, because they are our enemies, just as we are theirs.

forth from their tenuous shelter into the poisonous oxygen by a thoughtless toothpick or a casual floss.

Biologists are still arguing about some of the effects of early oxygen production. These days we believe that oxygen was something of a double-edged sword, clearing a way for our kind of aerobic life but causing some fearful problems for new life forms as well. Of course, it all happened so long ago that we are reduced to searching for hints to find clues to fuel inferences to feed hypotheses. It is all a bit tenuous.

On the other hand, we "modern humans" may have been directly shaped by an interaction with poisons. Around 1.8 million years ago, give or take a bit, something acted to change the line of curious bipedal apes that became us, and that something may have been poison.

In today's hunter-gatherer societies, meat is a chancy thing that provides a feast if the male hunters are lucky, while the female gatherers' daily food collection ensures survival.

Without straying too far, our female ancestors and their young could bring in regular basic food like tubers and small animals, but many tubers are poisonous. If early members of *Homo* were able to use fire, the secret to their survival may have lain in tubers that could now be disarmed by cooking. This would provide better daily nutrition for growing apelings who were changing into humanlings, as they were guarded by both parents, acting as partners.

Whether we were founded, shaped, sculpted, and formed by poison or not, there is a fascination with poisons that draws scientists and medical practitioners to them, either to use them or to study them, because in poison there is a hidden and secret power. The poisoner may be a figure of fear and evil, but the poisoner is the one who has the power of life and death—and this power is impossible to argue with.

How can we mere humans hope to compete with poisoners like our very remote ancestors? Perhaps we can't, but over time, we have done fairly well, as Pliny noted almost two millennia ago:

> . . . we rest not contented with natural poisons, but betake ourselves to many mixtures and compositions artificial, made even with our own hands. But what say you to this? Are not men themselves mere poisons by nature? For these slanderers and backbiters in the world, what do they else but launch poison out of their black tongues, like hideous serpents?
>
> Pliny, *The Natural History*, *c.* AD 70

Pliny was asphyxiated by Mount Vesuvius's volcanic gases in AD 79, and he was probably lucky to have made it that far without being poisoned by a rival, a ruler, or a relative. When you look back over the past few hundred years, however, not much has changed. The amazing thing is not the amount of poison around, it is that we manage to avoid most of these poisons. We have found ways to treat cassava and cycad seeds, to avoid the fish that will be bad for us, while at the same time identifying those poisons that might in small doses be good for something helpful, rather than harmful.

When rats encounter a novel food, they nibble a small amount, and if they feel ill effects, they do not try that food again. That is easy enough to explain as a response, but how do you account for people learning to treat poisonous food by soaking it, drying it, and soaking it again? Humans are able to communicate what they learn, which is important, but you have to wonder how many died before we discovered how to beat the poisons that had evolved to stop us eating so many potential foods.

We humans are lucky that a few inquisitive spirits can perform the tests and share their knowledge with many. William Buckland, nineteenth-century geologist and divine, and one of science's true eccentrics, took risks, but they were curiosity-driven rather than motivated by a love of humanity. Buckland pursued exotic foods to such an extent that he boasted about having eaten his way through much of the animal kingdom. According to the English writer Augustus Hare, there was even a moment of anthropophagy, or perhaps regophagy:

> Talk of strange relics led to mention of the heart of a French king pre-served at Nuneham in a silver casket. Dr. Buckland, whilst looking at it, exclaimed, "I have eaten many strange things, but have never eaten the heart of a king before," and, before anyone could hinder him, he had gobbled it up, and the precious relic was lost for ever.

Buckland, like some of the medical researchers we will meet later, was prepared to risk poisoning himself, but there were many, some of whom we shall meet now, who preferred to poison others.

2

A SLEW OF POISONERS

But in the central part of England there was surely some security for the existence even of a wife not beloved, in the laws of the land, and the manners of the age. Murder was not tolerated, servants were not slaves, and neither poison nor sleeping potions to be procured, like rhubarb, from every druggist.

Jane Austen, *Northanger Abbey*, 1818

If a few Locustas had flourished in Jane Austen's England, things might have been different, for many could perceive a need. The same 1859 issue of the London *Times* that reviews Charles Darwin's *Origin of Species* has a diverting tale from the divorce courts, giving one a feeling for the hoops two people had to jump through in order to obtain a divorce. A parade of witnesses described how they had seen named parties cavorting in a park in a state of partial undress, building a picture from which forensic minds would draw just one conclusion: naughtiness had happened in the bushes. In Victorian England, it required either flagrant adultery, a steely determination to be rid of one's spouse, or a conspiracy between both parties to achieve the sort of evidence required by the law—and the law was grimly determined to punish any hint of collusion.

In those days before humane divorce laws, this rigidity of the law all too often meant resort to the poison bottle. Cliffs were there for falling off and baths for drowning in, but poison offered more secrecy and, with luck, a death that might be mistaken for any one of a dozen natural causes.

This phenomenon was not unique to the northern hemisphere. Alfred Russel Wallace, Darwin's peer, came across similar behavior in his travels. Here is his account of what he saw and heard in Dili, in what was then Portuguese Timor:

> While I was there it was generally asserted and believed in the place, that two officers had poisoned the husbands of women with whom they were carrying on intrigues, and with whom they immediately cohabited on the death of their rivals. Yet no one ever thought for a moment of showing disapprobation of the crime, or even of considering it a crime at all, the husbands in question being low half-castes, who of course ought to make way for the pleasures of their superiors.
>
> Alfred Russel Wallace, *The Malay Archipelago*, 1869

No spouse killers can compete with the adulterating grocers, evil brewers, and toxic industrialists we will meet up with later, for they killed too few. But if a callous murderer lacks charm, the killers of spouses and other relatives for gain still have a certain Gothic curiosity, even a sort of charm. Two men, each both forger and poisoner in turn, exhibited a fatal charm for some of those whose paths they crossed. Each started in London and ended up in Australia: one of them was caught in a plot with more turnings than anything Charles Dickens ever planned, and Dickens met the other in jail.

Many Australians today are quite pleased when they find a convict in their lineage, secure in the belief that all convicts were transported by a draconian state for stealing a loaf of bread to feed their starving children, or for political offenses. There were a few genteel convicts, but they were very few. The felons of Australia were mostly felons through and through, desperate and bloodthirsty villains who managed to cheat the gallows by pleading to a lesser offense, accepting transportation to seek their betterment and the betterment of a penal colony on the other side of the world (and, as an afterthought, an intact neck). It was a good system, one that had provided workers for the sugar colonies of the West Indies and the Americas, and now it was Australia's turn. But while there were a few innocent early union organizers and feeders of the starving, most of the well-bred convicts were monsters like Tawell and Wainewright.

John Tawell was a charming, if self-centered, social climber who wanted desperately to be accepted in the Society of Friends, the Quakers, but they were intelligent people who could see through Mr. Tawell's pretensions. So, while working as a traveling salesman for a firm of London druggists, Tawell affected the plain clothing known as Quaker garb and hung around the fringes of the Society.

Not all apples in the barrel are of equal quality, and Tawell came to know a Quaker linen draper named Hunton who would eventually swing from the gallows after a career as a forger. Undeterred, inspired even, Tawell set out on the same route, seeking to gain possession of plates that would let him print bank notes on Smith's Bank at Uxbridge. Luckily for him, Smith's was a Quaker bank, and Quakers did not believe in capital punishment, so when he was caught a deal was cut and Tawell set off to begin 14 years in the colonies.

Being in a penal colony was to be in an open jail, where convicts lived and worked like any other colonist. For a clever convict, it could be the opportunity to make a fresh start and earn a fortune, and Tawell gave it a try—soon he was once again associating with the Friends, and no doubt claiming his willingness to atone. All the same, somebody who had needed to marry a pregnant housemaid in a hurry in London, and who did so outside the Society, who had been convicted of trying to defraud a bank owned by a Friend, and who now kept a mistress in Sydney would have needed to be fairly convincing.

In 1820 Tawell opened a pharmacy in Sydney, stating somewhat economically with regard to the truth that his "qualifications are neither unprofessional or irregular," and he prospered. Tidings of this new wealth reached his wife, who was still living in poverty in London. She demanded that the family be reunited, as was allowed by the benevolent state, and, in due course, she arrived in Sydney, which meant the end of the mistress. Around this time, Tawell gained a certain notoriety by ostentatiously destroying barrels of rum, pouring their contents into Sydney Cove. Then, to mark the visit to Sydney of two prominent Friends, James Backhouse and George Walker, he donated a plot of land that he said was to become the Friends' first meeting-house. (In fact, the gift was never completed, and after Tawell was hanged the land was sold and ended up as the site of a synagogue.)

Time and good fortune had made Tawell rich, but his wife was ill, so as his sentence had expired, he was allowed to escort her back to England. Her condition worsened, and he took on a nurse, Sarah Lawrence. When his wife died, Tawell made Sarah his mistress and had two children by her. In 1841, he married a Quaker widow, Eliza Cutforth, although annoying the Society of Friends by again being married outside the Society. Still affecting Quaker garb and Quaker virtues, he nonetheless continued to visit Sarah Hart, as she now called herself, at Salt Hill near Slough, only a day's excursion from London thanks to the new railway. This was also the year that Slough was linked to London by the even newer telegraph, designed by the famous Mr. Wheatstone and Mr. Cooke, and based on the electrical relay of the unknown Dr. Edward Davy, who had suddenly abandoned his wife in London and moved to Australia, leaving his father to sell his patents, and avoiding the need to seek a divorce.

By 1843, however, things were looking bad for Tawell. The Australian economy had nosedived, and his fortune, largely tied up in property in Australia, was in danger. When this grim news reached England, he saw that it was time to tighten the purse strings. Sarah would have to be let go, but as she would probably make a fuss, Tawell decided to sever the relationship once and for all. His first attempt to kill her, with morphia in her porter, made her ill but not dead, so on New Year's Day, 1845, he bought two drams of Scheele's Prussic Acid. He told the supplier that it was to "dress varicose veins."

He gave Sarah the cyanide in a bottle of stout. Then, alarmed by her agonized cries, he fled. He was, of course, wearing his distinctive Quaker garb, and a neighbor saw him heading for the railway station. Word reached the police, who rushed to the railway station, but the bird had flown on a slow train to London. Even a slow train could outrun the fleetest police constable, and once he reached London the villain would disappear. Still, his clothing made him distinctive, and no train could outrun the electric telegraph.

Somebody, history does not say who, thought to use the Wheatstone telegraph in a preview of the way Dr. Crippen would later be tracked and brought to justice by radio (see below). Apparently, the primitive telegraph had only 20 letters: it was missing C, J, Q, U, X, and Z, and tradition has it that the reference to "KWAKER" brought repeated interruptions from the London end, until a young clerk recommended that the sender be allowed to finish.

This may make a good story, even if words like "quick" and "queen" must have been sent previously without difficulty. The key issue is that London knew in sufficient time that a suspected murderer was on his way, and a policeman was waiting for him at the station. Tawell was followed to his home, arrested, and bound over for trial. During his trial there was a small sensation, thanks to his counsel, Sir George Edward Fitzroy Kelly.

It began with the well-known lawyer's first dramatic word: "Applepips," he said, letting it hang in the air. The seed planted in the jury's collective mind, he went on to describe how the post-mortem examination had found apple pips in Sarah Hart's stomach, and how she had received a sack of apples as a Christmas present from a friend. Being partial to the fruit, she had eaten a good deal of them in the days before her death. That, he suggested, was the origin of the prussic acid: it came from the apple pips! "In no other substance in nature, excepting bitter almonds, do you find such a concentration of the poison," he told them.

It was an ingenious, if desperate argument, but it didn't have the desired effect: the learned counsel was known ever after as "Applepip" Kelly, and orchards, markets, and greengrocer shops were left with piles of unsold apples, because the public now had a low opinion of the fruit. The jury had an equally low opinion of the defense and Tawell swung high for his crimes, at last. People were impressed by the telegraph, but not by Australia, which they saw as a place from which their surplus ruffians might return. Perhaps it was Tawell's case that gave Charles Dickens a plot line for *Great Expectations*, where Abel Magwitch is also a returned convict.

If he was a model for Magwitch, Tawell probably had even more in common with Wainewright the forger. This villain, otherwise known as Wainewright the poisoner, was only ever convicted of forgery—for which crime he was transported to Van

Diemen's Land—but few doubt his greater guilt. Wainewright holds a special place in the annals of poisoning because he was the inspiration for the villainous Slinkton in the Dickens short story "Hunted Down," and the wicked Varney in Lytton's *Lucretia*, as well as being the subject of a long biographical piece by Oscar Wilde.

In 1822, Thomas Wainewright was 28 years old, an accomplished artist who mixed with all the best people, like Byron and Coleridge. Being poor, and married, he used his artistic talents to criminal ends. He forged the signatures on documents to gain immediate access to some of his inheritance, which was held in a trust fund, but the money did not last. So in 1824, he forged his way to the whole of a legacy of £5,250. With his habits, however, this money was soon gone, and he began borrowing money from loan sharks and friends, running up large debts.

This is the point at which forger seems to have become poisoner, shortly after he moved with his wife and son into Linden House. This was the impressive country home where he had grown up, but which was now owned by his uncle, Thomas Griffiths. Within a year, Griffiths died under mysterious circumstances after convulsions quite similar to those caused by strychnine. The house and estate were left to Wainewright. Now he had a base from which to prey further, and lacked only suitable quarry.

So Mr. Wainewright invited his wife's younger sister, Helen Abercromby (as her name later appeared in law reports), and her mother to come and live with them. In no time, he had insured Helen's life with five different companies, lying about her age and the Wainewrights' financial situation. At this point, we are forced to draw inferences, but it is likely Helen was fed antimony to make her sick, and then strychnine in a restorative jelly, using sugar to disguise the bitter taste. She certainly died with convulsions.

It seems Helen's mother did not approve of the insurance

policies at all, but she had already died, with similar symptoms, leaving the coast clear. Now there was just one problem: the insurance companies were refusing to pay up! Wainewright started legal action to gain the payout, but then thought better of it, and ducked off to France for five years. On his return, however, he was recognized and arrested for the earlier forgeries, and shipped out to Van Diemen's Land. He would spend the rest of his life there, progressing from road building in a chain gang to working as a hospital orderly who was allowed to paint portraits of many local dignitaries and their families.

Oscar Wilde tells us how Dickens met Wainewright in June 1837, while visiting the jail where he was held. There are reminders here, too, of David Copperfield's visit to Uriah Heep in jail.

> While he was in gaol, Dickens, Macready, and Hablot Browne came across him by chance. They had been going over the prisons of London, searching for artistic effects, and in Newgate they suddenly caught sight of Wainewright. He met them with a defiant stare, Forster tells us, but Macready was "horrified to recognise a man familiarly known to him in former years, and at whose table he had dined."
>
> Oscar Wilde, *Intentions*, 1891

Our best authority for Wainewright's use of strychnine is once again Wilde, who tells us of the beautiful rings that served to show off the artist's delicate ivory hands, in one of which he used to carry crystals of *nux vomica*, or strychnine. And Wilde quotes De Quincey as saying that Wainewright's murders were "more than were ever made known judicially." In the end, though, Wainewright escaped a murderer's death. According to Wilde, Mrs. Wainewright was innocent of any involvement in the poisonings, and the Wainewrights seem to have parted company when he fled to Paris.

There is some doubt about whether Dickens drew on Wainewright in building Slinkton, and some critics think he used some aspects of William Palmer as well. Palmer was convicted, a few say wrongly, of murdering John Parsons Cook, though others believe he may have killed up to a dozen people in the comfortable belief that strychnine, given carefully—and Palmer was a careful doctor—would not be detectable.

The colony of Van Diemen's Land later changed its name to Tasmania to avoid the odium the old name carried with it. Palmer was known widely as "the Rugely murderer," and that town's burghers petitioned the prime minister to be allowed to change the name, saying Palmer had brought shame to them all. Legend has it that the prime minister agreed, so long as they named the town after him, so the burghers of Rugely thanked the prime minister and bothered Lord Palmerston no more.

Palmer was convicted and hanged for the murder of John Cook, a wealthy horse-racing associate, but the facts of the case gave rise to such suspicions that his dead wife's body was exhumed and examined. Traces of antimony were found, and most people drew the not unreasonable conclusion that murder most foul had been done to her as well. After all, they told each other, in April 1854 Palmer had insured his wife, Annie, for £13,000 and paid just one premium of £760. Then, after going to a concert in a light dress, she was thought to have caught a chill. After taking some tea with sugar and no milk, and toast, she began vomiting.

Her family physician, the elderly Dr. Bamford, took this illness to be English cholera and prescribed calomel, colocynth, and an "opening drought," as a laxative was then called. Witnesses said later that Dr. Palmer ordered his wife to be given just a small amount of prussic acid, to reduce her retching. She died in spite of or thanks to the treatment, and Dr. Bamford and his colleague Dr. Knight, elderly, deaf, and Annie's guardian, certified English cholera as the cause of death. Bamford, however, had relied on Palmer's description of the symptoms.

Professor Alfred Taylor, who became rather obsessed with the case, said Annie Palmer's death was caused by doses of antimony, "the solid sulphide of this metal was found in the stomach after death, while the metal itself pervaded the whole of the tissues. A respectable physician, with only a superficial knowledge of the real facts of the case, wrote a pamphlet to prove this woman had died of cholera." Taylor went on to observe, rather caustically, that had such people been treating her case, they would never have realized that poison was involved.

Taylor had given evidence at Palmer's trial for the murder of Cook, testimony that lasted almost a day. He said he had originally diagnosed antimony as the cause of death because he found a half grain of it, a comparatively tiny amount. Later, when he heard that Palmer had bought strychnine, and that Cook had suffered convulsions, he changed his mind. According to Palmer's supporters, Cook's brain showed no signs of the damage that ought to be caused by strychnine. More importantly, they said, there was a delay of an hour and a half between the onset of symptoms and the time when Cook was supposedly given the poison, and this was too long. Tetanus, they said, was what killed Cook. Like at most major trials, there were plenty of lay observers with a keen interest and strong views about the case. Robert Graves quotes one of these:

> In antimony, though great his faith,
> The quantity found being small,

Taylor's faith in strychnine was yet greater,
For of that he found nothing at all.

Robert Graves, *They Hanged My Saintly Billy*, 1957

From a poisoner's viewpoint, the best poison needs to be tasteless, impossible to detect in a corpse, and have symptoms that are like those of a known illness. All that is then needed is a suitable outbreak of the illness in the vicinity. And who knows most about poisons? Why, a doctor, of course!

Sherlock Holmes at his work bench, a Sidney Paget
illustration from the *Strand* magazine.

Perhaps this is what Arthur Conan Doyle had in mind when he has Sherlock Holmes say in *The Speckled Band*:

> Subtle enough and horrible enough. When a doctor does go wrong he
> is the first of criminals. He has nerve and he has knowledge. Palmer
> and Pritchard were among the heads of their profession.

Never mind the Hippocratic oath, with its promise neither to administer a poison to anybody when asked to do so, nor to suggest such a course—Edward Pritchard regarded poison as an easy way out. His last way out came at the end of a hangman's rope in Glasgow on July 28, 1865, the last person publicly executed in Scotland. He drew a good crowd, around 100,000.

Pritchard had qualified as a doctor and gained a commission as naval assistant surgeon in 1846. In 1850, he married Mary Jane Taylor, and in 1851 resigned from the Royal Navy and took up a position as a GP in Yorkshire, but moved to a new practice in Glasgow in 1860. In 1863 their house was badly damaged in a fire that killed (perhaps conveniently) a servant girl. Not long after, Pritchard made their 15-year-old maid pregnant and promised to marry her if his wife died.

The girl then agreed to his performing an abortion on her, at which point Mary Pritchard conveniently became seriously ill, on February 1, 1865. Her elderly mother came to nurse her but was under Pritchard's roof for only one night before she fell ill as well, and died on February 25. Mary died on March 18, and Pritchard had the coffin unscrewed so he could kiss the lips of the wife he had poisoned, for which act he would later be labeled "the Human Crocodile."

Pritchard's undoing came when the Procurator Fiscal received an anonymous letter, alleging that the doctor had poisoned both women. The Procurator Fiscal is a Scottish law officer, whose position still exists—indeed, the office even has its own Web site. Essentially, the role is that of prosecutor, but one with the power to direct the police—just the sort of person Pritchard would not want to have nosing around.

The bodies were exhumed, enough antimony was found to justify a verdict of murder, and Pritchard made his farewell public appearance on the gallows. He was, by all accounts, an evil-tempered man, and one who preyed on young girls, so Doyle chose a fine pair of examples for Holmes to cite.

One would-be author in recent times has argued that Doyle, himself a medical practitioner, could also have been named.

Rodger Garrick-Steele, a man described variously as a former psychologist, a writer, an author, an aspiring author, "a struggling author whose book has been rejected by 90 publishers" and "a former driving instructor," claimed in 2000 that Doyle had obtained the story of *The Hound of the Baskervilles* from Bertram Fletcher-Robinson, and that he then coerced Fletcher-Robinson's wife into poisoning her husband with laudanum. Garrick-Steele claims to have spent eleven years researching the story. He lived in the Fletcher-Robinsons' old home, and claimed that uneasy ghosts in the house alerted him to the affair.

When last heard of, Steele was still trying to hawk his story for a movie, but news had leaked out of his attempt to involve Scotland Yard by reporting a murder, while judiciously not mentioning that the alleged crime had taken place almost 100 years earlier. Most people feel that Dr. Doyle's reputation remains untarnished. The police were certainly unimpressed, and Chief Superintendent Brian Moore was quoted in the *Daily Express* as saying, "We get a lot of people who try to exploit our name for a variety of reasons and really we are far too busy to take further action and prosecute these people. However, I do look forward to the opportunity of having a further word with Mr. Garrick-Steele when I would be pleased to give him some friendly advice."

There have been many other doctor-poisoners. The story goes that, some years before 1881, Professor Robert Christison, who held the chair of Jurisprudence at Edinburgh University, was giving evidence in a court case. When he started to explain that there was only one substance that could not be traced by doctors, the judge interrupted him before he could name the substance, saying it was better that the public be kept ignorant of the name. Back in his lecture hall, however, Christison named the substance as aconitine—and among his students was one George Lamson.

In March 1882, George Henry Lamson was tried, and swiftly convicted and executed, for poisoning his brother-in-law. Percy Malcolm John was a paraplegic, and before he died Lamson was supposedly showing him how to put "sugar" in a gelatin capsule as a way of taking vile-tasting medicine. Lamson had bought two grains of English (Morson's) aconitine a few days earlier, and this, said the prosecution, was what was actually in the capsules. Within five minutes, the symptoms of aconitine poisoning began. Sadly for Lamson, forensic science had moved forward since his student days, the poison was identified—and he swung for his crime.

Dr. Couty de la Pommerais is generally more famous for something he did not do than for anything he did. He is supposed to have agreed to wink three times at a given signal after he was guillotined, if he found himself able to do so, but we will come back to this in a minute.

A certain Madame de Pauw (the only name we seem to have for her) was the widow of a friend of his. De la Pommerais first made her his mistress, then dropped her. She was 40 and in good health when her former lover suddenly took up with her again, and then invested in several excessively large policies on her life. When she died, there were no reasons to suspect foul play, so she was buried. When de la Pommerais lodged his insurance claims immediately after her death, her body was exhumed, but no signs of poison were found. Moreover, there were no records of her symptoms, but aqueous and alcoholic extracts of the corpse's stomach and intestines were prepared and applied to animals, the normal test in those days.

A rabbit and a dog died from these extracts, and an alcoholic extract of scrapings from the floor where she had vomited contained enough poison to kill another rabbit and another dog, while a scraping from another part of the floor had no effect. The conclusion was that death had been caused by a vegetable poison that could not then be identified.

The doctor had large amounts of a number of poisons in his possession, and while he had recently bought significant amounts of digitalin, very little of it was left. Now, the problem for de la Pommerais was that he was no normal medical man: he was a homeopathic practitioner, whose remedies consisted of remarkably dilute doses of poisons, taken in small amounts, as Oliver Wendell Holmes was keen to remind people.

The amount of digitalin was such, however, that it was hard for him to explain away its purchase and subsequent use for any reason other than criminal. This, together with his unexpected renewal of the relationship, and the insurance policies he had taken out, made a sufficient case to convict him. Add to the list a collection of forged letters and correspondence that he had prepared, showing that he had expected Mme de Pauw's sudden death, and his execution in 1864 is less than surprising.

What is surprising is the legend about whether the doctor could wink with a severed head. In fact, this appears to have been a popular nineteenth-century Parisian myth, one told of a number of guillotined people. In de la Pommerais' case, it began as a fictional work by Villiers de l'Isle-Adam—*The Secret of the Scaffold*, published in 1883. Alfred-Armand Velpeau was a celebrated surgeon who had his moment of fame when he gave his name to the Velpeau bandage, now largely unknown. In *The Secret of the Scaffold* Velpeau supposedly approached de la Pommerais in the condemned cell and persuaded him to try to wink three times at an agreed signal, in the name of their joint love of science. Velpeau had died in 1867, and so was in no position to confirm or deny anything, but on his behalf, let me say that the meeting never happened. Besides, it is unlikely anybody would care much about de la Pommerais, who simply had it coming to him.

Dr. Hawley Harvey Crippen was another homeopath gone wrong, but in his case we might perhaps be tempted to have just a tinge of sympathy for him and his Pugh-like mustache. He was 50, and his wife was a domineering woman, a fringe performer in the music halls, where she used the name Belle Elmore. The Crippens had married in their native United States when she was 19 and he a widower of 30. She was for the high life, a disappointed diva who never made it, he a would-be quiet homebody and operetta-goer.

The young Belle had been impressed by Crippen's MD, but America had lost faith in homeopathy, and so they moved to London. There, Belle flirted as actresses could and did, and then in 1901, Crippen met Ethel Le Neve. Like his wife—who was born Kunigunde Mackamotzki and also used the names "Cora Motzki" and "Cora Turner"—Ethel had changed her name from the plain "Neve" to the more stylish "Le Neve." As the Crippen marriage stalled in irritation, Ethel came to love him. Then, after he found Belle in bed with another man, Crippen and Ethel became lovers.

For whatever reason, the Crippens did not do the sensible thing and arrange a divorce. Instead, Crippen bought five grains of hyoscine hydrobromide, now used mainly as the active ingredient in a well-known remedy for motion sickness. Then, one night in 1910, Crippen poisoned his wife, cut her rather ample body into smaller parts, and disposed of most of the bits. Failing to dispose of all of the bits was to be a fatal error.

Crippen also made the mistake of allowing Ethel to wear one of Belle's brooches in public. It was a large and recognizable piece, one that her friends spotted immediately. He had first told them his wife had gone off to California, then that she had died there and been cremated, but they knew Belle was a Catholic and would not have been cremated. Thinking they detected a distinct odor of large rodent, Belle's friends went to the police. Alarmed, Crippen and Ethel disappeared, whereupon the police searched the Crippens' cellar and found human remains. The search was now on for the missing doctor and his lady friend, and the press reveled in what they called the "London Cellar Murder."

Crippen's most distinguishing feature was his mustache, so it had to go; Ethel disguised herself as a boy, and they sailed together for Canada. The captain of the SS *Montrose* had seen a newspaper report of the case, and he noticed that his passenger "Mr. Robinson" and his supposed son seemed overly affectionate toward each other. This was 1910, and his was a modern ship, so the captain sent a message back to London via his Marconi wireless telegraphy set: "Have strong suspicion that Crippen London Cellar Murderer and accomplice are amongst saloon passengers. Moustache shaved off, growing a beard. Accomplice dressed as a boy, voice, manner and build undoubtedly a girl."

To make certain, the cunning Captain Kendall then told "Mr. Robinson" a joke. There is an art of divination called geloscopy, where the diviner assesses a person by the way he or she laughs, but it is unlikely the captain was so much a geloscopist as a would-be amateur detective. The circulated description had identified the fugitive as having false teeth, so Kendall told Crippen a joke to make him laugh. Sure enough, the suspect's teeth were loose and clearly false.

To the world's press, it was Tawell all over again, and so it was, except that it was completely different. Just as Tawell's arrest had proved the value of the telegraph to two generations, now Crippen's capture proved the effectiveness of radio, but after that they diverged. This time, the ship sent the first message back to the police, and Chief Inspector Dew was sent in a faster ship, the SS *Laurentic*, to overtake the fugitives. Meanwhile, the captain's message was released to the press, so the English public knew all about the chase even as it was happening. Despite the Tawell parallels, there were other, more recent items in the locker of journalistic clichés waiting to find their way into the tale.

A fairly recent encounter that had captured the public's imagination was the meeting in Africa of Stanley and Livingstone, usually portrayed as a classic of British reserve. The true story was told in *Nature* in 1872, and reveals that it was Stanley's deference to the assumed reserve of the Arab traders who were looking on when they met, rather than his English reserve, that led him to behave as he did.

> In an instant, he recognised the European as none other than Dr Livingstone himself; and he was about to rush forward and embrace him, when the thought occurred he was in the presence of Arabs, who, being accustomed to conceal their feelings, were very likely to found their estimate of a man upon the manner in which he conceals his own. A dignified Arab chieftain stood by, and this confirmed Mr Stanley in his resolution to show no symptoms of his own rejoicing or excitement. Slowly advancing towards the great traveller, he bowed and said 'Dr Livingstone, I presume?' to which address the latter, who was fully equal to the occasion, simply smiled and replied 'Yes'. It was not till some hours afterwards, when alone together, seated on a goat skin, the two white men exchanged those congratulations which both were eager to express, and recounted their respective difficulties and adventures.

> *Nature*, 1872

So, fully aware that the eyes of the world were upon him and that a ringing phrase would carry him into the history books, Chief Inspector Dew later claimed that he boarded the *Montrose* and approached the fugitive with the words "Dr. Crippen, I presume?" The fugitive admitted his identity, expressed relief that it was all over, and was taken back to London. Incidentally, Dew may have been embroidering his version a little, as other accounts have him saying, "Good afternoon, Dr. Crippen, remember me? I'm Inspector Dew with Scotland Yard," but whatever he said, Crippen's flight from justice was ended.

Under questioning, Crippen offered yet another story. He now said Belle had run away with a lover, and that his earlier excuses had been to avoid a scandal, but Sir Bernard Spilsbury showed there was hyoscine in the remains in the cellar, and a small patch of skin bore a scar identical to one Belle was known to have had. Perhaps if Crippen had told his side of the story, he might have got off, but that would have involved Ethel, and he could not have that. So Dr. Crippen, as he will always be known, was hanged by the neck until dead.

Given that he remained silent to protect Ethel's reputation, perhaps we should not make too much of the theory that he administered the depressant hyoscine to Belle in good faith in order to curb her sexual appetite, but it all went wrong. Crippen may deserve some sympathy, but his name as a villain is all he has left, so we shall leave it to him.

I assure you that the most winning woman I ever knew was hanged for poisoning three little children for their insurance-money, and the most repellent man of my acquaintance is a philanthropist who has spent nearly a quarter of a million upon the London poor.

Arthur Conan Doyle, *The Sign of the Four*, 1890

One of the aspects of poison that has most excited the public's enthusiastic horror is the way in which people poison others in order to collect insurance money on them or to gain an inheritance. This was by no means a new repugnance: while ruling in Spain—before he became emperor in AD 68 and did away with Locusta—Galba sentenced to death a man who poisoned his ward to inherit a property. The form of death for a shameful crime was crucifixion, and when the man protested that he was a Roman citizen, Galba replied, "Let this citizen hang higher than the rest, and have his cross whitewashed."

An interesting example is the 1850 case of the then Count and Countess of Bocarmé. I need to stress the *then*, because there is a modern holder of the title with an apparently unblemished record, albeit with rather blemished predecessors. Count Hyppolite de Bocarmé was part Belgian and part Dutch, and he claimed that, having been born at sea in a wild storm, he was allowed to run wild as a child. In a fairy tale, such a person would either be the hero who kills the dragon or, more commonly, the villain.

This is just a common tale, and he was the villain. Worse, he was an almost penniless villain when he acquired his title and the château de Bitremont, near the Belgian community of Bury. The count did what all villains did under those circumstances, and married an heiress. Sadly, although he had developed an interest in science, he failed to do his research as well as he should have. His heiress was both a big spender like himself and not as well-equipped in the fortune department as he had thought. This left the improvident pair no choice but to find some other source of financial support.

Madame and her sickly brother, Gustave, inherited some money when their father died, and the count also sold some of the land around the château, but it was still not enough to maintain their lifestyle. By 1849, they were running out of ways to raise spending money, and they were pinning their hopes on Gustave's early death. Then came the bombshell: Gustave had bought himself a château of his own and, what was worse, had developed an interest in the former owner.

Soon word came that Gustave would arrive at noon on a certain day to announce his engagement. The plot began to thicken as a number of unusual plans were laid. For example, the children usually ate with the adults, but this day they were to eat in the kitchen. What was more, the meal was to be served by the countess herself, and not by the château servants.

The first the servants knew of any dark deeds was when they were called in to help Gustave, who, the Bocarmés claimed, must have had a stroke. Curiously, the count kept pouring vinegar down Gustave's throat, and washing him with vinegar. A suspicious mind might have dallied at least briefly with the thought that the count wished to hide something. The countess had all of Gustave's clothes removed, and threw them into hot soapy water, which might have intrigued the same suspicious mind a little, and she spent most of the night scrubbing the dining room floor. By this time, even the most naïve would begin to think something was not entirely as it should be.

Late that night, when the count and countess finally went to bed, it all boiled over below stairs. The servants decided to tell the local priest of what they had seen, but word of the events at the château de Bitremont also reached the local examining magistrate, who arrived to find a fireplace full of the remains of half-burned books and papers, and a dining room floor littered with wood shavings. Gustave's body showed signs of burns, which suggested a corrosive liquid had been applied, and sulfuric acid seemed a likely possibility.

Nicotiana tabacum

In reality, the poison used was nicotine, and it is possible the count's research was as out of date as Dr. Lamson's had been. Three years earlier, Mathieu Orfila, the leading French toxicologist, had publicly lamented the problems of identifying vegetable poisons, but this lament did not take into account the skill of a Belgian analytical chemist, Jean Stas—and neither did the count.

In short order, Stas ruled out sulfuric acid and succeeded in identifying the poison in Gustave's organs as nicotine. He suggested that the magistrate inquire as to whether anybody in the Bocarmé household had access to a preparation of nicotine. A gardener reported that he had helped the count prepare extracts of nicotine, and a search revealed the corpses of several cats and ducks that had been used as test animals. On top of this, police inquiries found a number of pharmacists who had been consulted by the count about extracting nicotine. Of course, Rule Number One for Great Poisoners is to do your research in disguise in an out-of-the-way library, under an assumed name—for that reason alone, the count is only ever likely to make the list of top ten failures.

To see why he was such a failure, we need to understand the chemistry that trapped him. Most organic chemicals are soluble either in water or in alcohol, or they may be insoluble in both—very few are soluble in both water and alcohol. The alkaloids, though, are the exception to the general rule. So if you take bodily remains containing an alkaloid, and soak them in alcohol to which an acid has been added, the alkaloid, and any alcohol-soluble body chemicals, will be dissolved. When this solution is evaporated down gently, and water is added to the resulting solid material, only the water-soluble alkaloid will dissolve, but this all hinges on there being acid present in the first place.

Good fortune was smiling on the officers of the law. Someone had added a dilute solution of what used to be called acetic acid, now better known as ethanoic acid, to the body. And who was this thoughtful person? Why, the count himself, who provided the preserving acid in the form of vinegar, which gets its distinctive sour taste from the mild acid produced when ethanol is oxidized to ethanoic acid. The count's ministrations and washing had preserved the evidence of his crime perfectly. In the ensuing trial, the countess claimed her husband had forced her into it, while he admitted to making the poison, which he stored in a wine bottle, but claimed his wife had actually dosed Gustave. In the end, he was executed and she was acquitted—and presumably gained Gustave's inheritance, since she had been found not guilty.

Frederick Henry Seddon was another foolish man whose own actions brought him down. He was an insurance agent from North London who had sold an annuity to his lodger, Eliza Mary Barrow, and then poisoned her. This saved him the cost of paying her the annuity. When prosecutor Rufus Isaacs asked him if he had liked Miss Barrow, Seddon evaded the question, which was perhaps the worst response for him to have made.

He was found guilty, the press said at the end of the trial, largely because his testimony was too obviously rehearsed, too precise in the trivial details. Miss Barrow had a history of ill health, but Seddon had too many details at his fingertips. Arsenic was found in Miss Barrow's remains, which it seems Seddon had extracted from flypapers. Seddon was careless enough to leave cuttings on the celebrated Maybrick case, detailing the effects of arsenic on the human body, in his rooms.

The Maybrick case is one that continues to draw interest,

because James Maybrick, the alleged victim, may (or may not) have been Jack the Ripper. This claim is based on the existence of a diary, "found" in 1992, and now generally dismissed as a hoax. In 1889 James Maybrick's wife, Florence, was convicted of his murder but, with hindsight, there seems to be good reason why she should have been acquitted.

Mrs. Maybrick was born Florence Elizabeth Chandler in 1863 but preferred to be called Florie. Florence Chandler was American, and her mother, Caroline, seems to have been what was known in Victorian times as an adventuress. The mother may or may not have poisoned her first husband, a lawyer, and also her second husband, a Confederate officer named Franklin Du Barry, who was conveniently buried at sea.

Caroline then married Baron Adolph von Roques, and thus became a baroness, a title that she traded on when she left him. In 1880, she and Florie sailed from New York to Liverpool on the *Baltic*, and Caroline set her sights on James Maybrick. He preferred Florie, despite or perhaps because of her being 24 years his junior. They were married in 1881.

Maybrick was an arsenic and strychnine eater, and had been one since 1877, when he started dosing himself as a remedy against malaria. The Maybrick family was well-off, and James's brother Michael, under the pen name Stephen Adams, was a famous songwriter of such ditties as "The Holy City," a great seller in sheet music. It remains a popular Victorian religious ballad to this day, and it was one of perhaps a hundred songs from the prolific Mr. Adams.

Neither Michael nor James Maybrick appears to have been what we might call a person of orthodox church- or state-approved sexual orientation and interests. This is not to suggest that we sit in judgment on them, but it is worth keeping in mind when we consider the righteous wrath of Florie's judge when he denounced her own peccadilloes. In other words, James was undoubtedly unfaithful to Florie on many occasions, but in those days, that was regarded as the man's right. Even so, while he may well have known London prostitutes, it seems unlikely he killed any of them.

The evidence against Florence Maybrick seems to come from a variety of biased sources, some of them worked up after James's death. She undoubtedly had a lover, a cotton broker named Alfred Brierly, and her husband's family was aware of a letter she had written him. Whatever she said in the letter, it was enough to set Michael Maybrick against her, and he did everything in his power to see her hang. He also either had James change his will, or himself forged a new one that cut Florie out of any significant inheritance.

By a curious coincidence, Florence's judge, Mr. Justice Stephen, was the father of J. K. Stephen, another suspect for Ripperhood. J. K. Stephen was the tutor of Queen Victoria's grandson, Prince Albert Victor Edward, yet another possibility. J. K. Stephen was an excellent light poet, and his elder brother, Sir Leslie, was the father of Virginia Woolf and Vanessa, who later married Clive Bell. It was a talented but unstable family, and might, perhaps, have produced a Jack, but there are many candidates for the role, including a Dr. Cream, who was a medical student with Arthur Conan Doyle. He was later hanged for poisoning prostitutes by giving them strychnine capsules, saying the capsules would improve their complexion. As Cream began his drop to oblivion, he is said to have shouted, "I am Jack . . ."

The fact that Cream was in jail in America when at least one of the Ripper murders he was supposed to have committed took place is of no more account than the unbreakable alibis of Lewis Carroll. His candidature was based on curious anagrams hidden in his verse. Carroll was in fact a brilliant mathematician and a constructor of clever logical puzzles, so he was unlikely to have constructed such clumsy, error-ridden clues as have been put forth.

The prosecution claimed that Florence had used arsenic, extracted from flypaper, to poison her husband. In her trial, Florence admitted that she had indeed soaked flypapers to get arsenic, but stated that she used the arsenic for her complexion. She also said that James had asked her to add some of "my powder" to meat juice that he was given while ill. It is reasonable to assume that this was arsenic or strychnine, and arsenic was certainly found in the meat juice, but the judge declared to the jury that Florence was a liar, and no mention of James's arsenic and strychnine habits ever came out in the trial.

One thing that did come out was that Florie and James had argued about her infidelity (women weren't supposed to be players back then) shortly before he died. Her infidelity alone was almost enough to convict her for a judge who loathed her and was himself going insane at the time.

Florence may have been convicted, but due either to pressure from America or evidence that could not be made public at the time, her death sentence was commuted to life imprisonment. She would never be paroled while Queen Victoria lived, as Her Majesty did not approve of her or her alleged crime. In fact, Florence would spend 15 years in jail before being quietly released in 1904. She left Britain and settled in the United States, where she wrote an autobiography, *My Lost Fifteen Years*, protesting her innocence. She died in obscurity in 1941, aged 79, her identity known only to a handful of neighbors.

Madeleine Smith's alleged victim was another arsenic eater, and, like Florence Maybrick, Madeleine would live out her life quietly in America. There is no doubt that her lover, Emile L'Angelier, died of arsenic poisoning in 1857, but there are some peculiar aspects to the case that make the Scottish verdict of "Not Proven" seem quite appropriate: on the balance of probabilities, it seems likely Emile killed himself as an act of revenge against the woman he loved, the woman he was determined no other man should have, the woman who had ended their relationship.

Madeleine was 22 when she stood trial in 1857, but as far back as 1852 Emile had boasted that he was an arsenic eater. In 1856, Madeleine decided to break off her relationship with Emile, as her family was implacably opposed to it. For some reason, she sent the houseboy out to get cyanide, perhaps because Emile had urged her to do so, but the boy came back without it.

Madeleine's supporters argue that when she announced she was breaking with him, Emile probably decided to feign being poisoned out of jealousy. He wanted to make sure she never married another and went to great lengths to tell everyone who would listen that Madeleine was trying to poison him, even claiming to have been ill after drinking coffee and chocolate she had prepared. At Madeleine's trial, Emile was quoted as saying, "I can't think why I was so unwell after getting that coffee and chocolate from her," and "It is a perfect fascination, my attachment to that girl; if she were to poison me, I would forgive her."

Her purchase of poisons, combined with Emile's comments, could have been enough to destroy Madeleine. She claimed that Emile had urged her to buy the poison so they could take it together, but the real cornerstone to Madeleine's defense was that the earliest purchase of arsenic the prosecution could discover was on February 21, two days after Emile was first supposed to be "poisoned." Emile had reason to believe she had already obtained some poison by February 19, so it would be easy for him to make the error but remarkably hard for her to administer the first dose.

More to the point, the final dose taken by or fed to Emile was more than half an ounce, an amount that could not be missed if he was being poisoned by some other person. By now, Madeleine had bought a second batch of arsenic, again at Emile's request say her supporters. As required by law, the arsenic Madeleine bought was stained. One pharmacist used soot, and another indigo. The arsenic found in Emile's stomach was white—wherever it came from, it was not from the supplies she bought.

Despite a lack of concrete evidence, Madeleine was lucky to be tried in Scotland, where a verdict of "Not Proven" stopped short of an outright acquittal but, nonetheless, allowed the defendant to go free. She would possibly not have been as lucky in an English court.

Many chose freely to partake of poisons as part of their lifestyle rather than as part of their deathstyle. Thomas De Quincey would make the practice famous with his sensational and serialized *Confessions of an Opium Eater*, though his friend Coleridge would describe De Quincey as a mere dilettante in comparison to himself when it came down to the amount of laudanum consumed.

In time, people would become more aware of what they were doing to their bodies—not that a greater knowledge of those poisons we call drugs seems to change people's behavior much. Ironically, there was no need to be a deliberate opium eater, or to partake of large quantities of arsenic or strychnine, when so many poisons were already in the everyday foods people ate.

3

POISON AND FOOD

"I've had 19 straight whiskies. I believe that's the record."

Dylan Thomas's reputed last words, 1953

We talk happily of "intoxication" and "toxicology," usually without making the link between the words, even if we have heard of "alcohol poisoning." Alcohol *is* a poison, though one we have learned to deal with. We turn an enzyme called alcohol dehydrogenase on it, a protein that rips the alcohol molecule apart, taking away its poisonous power.

We can break down alcohol because bacteria in our intestines have been producing small quantities of alcohol over a very long time. On average, each day your stomach is delivered the alcoholic equivalent of half a pint of beer, and as this small dose seeps into your bloodstream, the enzymes attack it and render it powerless, stopping your otherwise inevitable internal pickling (and mine as well, for you are not alone in this). It is only a slow production of alcohol, though, which is probably just as well or the government would want to slap a tax on it. This bodily process explains why we can intoxicate ourselves with falling-down juice and still be conscious enough the next day to wish we were dead.

We have a secondary defense against alcohol; the ability to eject the contents of our stomachs when we need to get rid of something disagreeable. It's an ability we share with our canine companions but one which we humans seem able to wear down with experience. Hardened drinkers such as the late Dylan Thomas can withstand the increase in acetaldehyde produced as the enzymes in their liver break down their overload of ethanol. In the end, the alcohol will kill them without being summarily thrown up and out.

People jokingly refer to alcohol as a poison, but in reality there is enough ethanol in a liter of spirits to kill most adults. Not everyone has the same form of the alcohol dehydrogenase enzyme. Native Americans commonly have a less efficient form of the enzyme, leaving them more susceptible to the intoxicating effects of liquor than others—the stereotypical "drunken Indian" of mid-twentieth-century cowboy movies was as much a victim of his genes as he was of the rotgut, or those who sold it.

More commonly, though, the usage is figurative. In *The Tenant of Wildfell Hall*, for example, Lord Lowborough declares he will give up alcohol: "'It's rank poison,' said he, grasping the bottle by the neck, 'and I forswear it.'" Some alcohols are more poisonous than others. Take Ginger Jake, for example, a popular substitute for liquor during Prohibition in the United States. This was an alcoholic extract of Jamaica ginger, and legally listed in the *U.S. Pharmacopoeia* as a cure for assorted ailments. It tasted so horrible that the authorities thought it would surely be safe enough to sell, but the poor bought it anyhow to satisfy their need for a buzz. Sadly, in 1930 one batch was accidentally adulterated with poisonous tri-orthocresyl phosphate. As many as 50,000 people were affected, their symptoms beginning with cramps and sore calf muscles but developing into a form of leg paralysis known and celebrated in song as Jake Leg.

Alcohol can sometimes make another substance even more dangerous. Some otherwise harmless fungi are toxic when ingested with alcohol—*Morchella* and *Coprinus* are two of them. Later we will look at some of the problems caused by various foreign substances in rum, beer, wine, and cider, but let us first consider other toxic beverages.

Benjamin Thompson is best known to physicists for his studies on the conversion of mechanical energy to heat, observed during the boring of cannon barrels, but this unusual polymath was never slow to label something a poison. An American self-made scientist and probable English spy who ended up as the German Count Rumford, Thompson denounced beer as a poison that was debilitating the German workers. He believed they would be better off drinking coffee, but he knew something had to be done about people spoiling the coffee by boiling it, so he invented the percolator. He might have been in favor of coffee, but he had severe reservations about tea:

> When tea is mixed with a sufficient quantity of sugar and good cream; when it is taken with a large quantity of bread and butter, or with toast and boiled eggs; and above all, *when it is not drank too hot*, it is certainly less unwholesome; but a simple infusion of this drug, drank boiling hot, as the Poor usually take it, is certainly a poison which, though it is sometimes slow in its operation, never fails to produce very fatal effects, even in the strongest constitution, where the free use of it is continued for a considerable length of time.
>
> Benjamin Thompson, Count Rumford,
> *Of Food and Particularly of Feeding the Poor*, 1796

Thompson also offered a "Receipt for making brown soup." Note the use of rye in the recipe, and his warning about copper poisoning, both of which we shall come across later.

> Take a small piece of butter and put it over the fire in a clean frying-pan made of iron (not copper, for that metal used for this purpose would be poisonous);—put to it a few spoonfuls of wheat or rye meal; stir the whole about briskly with a broad wooden spoon, or rather knife, with a broad and thin edge, till the butter has disappeared, and the meal is uniformly of a deep brown colour; great care being taken, by stirring it continually, to prevent the meal from being burned to the pan.
>
> Benjamin Thompson, Count Rumford,
> *Of Food and Particularly of Feeding the Poor*, 1796

If G. K. Chesterton is to be relied on, it was the grocers, the people responsible for selling goods such as tea in bulk, who were the greatest poisoners, because of the adulterations they committed on the food they sold.

> He sells us sands of Araby
> As sugar for cash down;
> He sweeps his shop and sells the dust
> The purest salt in town,
> He crams with cans of poisoned meat
> The subjects of the King,
> And when they die by thousands
> Why, he laughs like anything.
>
> G. K. Chesterton, "Song Against Grocers," 1914

As early as 1612, the grocers themselves were complaining of others doing the same thing. The Master and Wardens of the Grocers' Company of London were enraged about cheap foreign competition. They complained that "a filthy and unwholesome baggage composition was being brought into this realm as Tryacle of Genoa, made only of the rotten garble and refuse outcast of all kinds of spices and drugs, hand overhead with a little filthy molasses and tarre to work it up withal." (Remember this "Tryacle" or treacle, because we will return to it in the next chapter.)

Take tea, for example: it was normal to find sulfate of iron in tea and beer; and ferric ferrocyanide, calcium sulfate, and turmeric would as like as not be lurking in Chinese tea. Spent tea leaves were sometimes refurbished, the appearance of the leaves being improved by a process called facing. This involved the agitation of the used leaves with soapstone and Prussian blue. Some of these substances might not strictly be poisons, but they weren't exactly good for you either.

In December 1859, six people in Clifton, with no obvious connection to each other, suffered from the usual symptoms of poisoning by arsenic. Investigations revealed the source to be Bath buns. A local confectioner had used, or so he believed, chrome yellow (lead chromate), to give his buns a rich yellow color and make them more attractive, but the druggist had supplied him with orpiment or arsenic sulfide instead.

Lead chromate would not be a desirable additive from today's perspective, but at least it was not absorbed, which was probably just as well, since it was commonly added to both mustard and snuff. Red lead was also added to snuff, and was responsible for Gloucester cheese's nice red hue.

In truth, almost everybody was adulterating food, slipping "extras" into food and drink, or passing the extras off as food and drink. In one case, early in the twentieth century, an alleged cider was claimed to be prepared from concentrated apple juice. On analysis, the "cider" turned out to be sugar, fruit essence, and aniline dye, with not even a trace of apple juice.

The Black Book: An Exposition of Abuses in Church and State was published in London in 1832, and Michael Gilbert quotes the following alarming report:

> We had a singular instance of this in the case of Mr. Abbott, brewer and magistrate, of Canterbury. This man had for a long time been selling, according to Lord Brougham's statement, rank poison in the beverage of the people. It appears he had been selling a beverage resembling beer, manufactured from beer-grounds, distillers' spent wash, quassia, opium, guinea pepper, vitriol, and other deleterious and poisonous ingredients. The officers of Excise having examined this worthy magistrate's premises, found 12 lbs. of prepared powder, and 14 lbs. of vitriol or copperas; in boxes; which, if full, would have contained 56 lbs.
>
> *The Black Book*, 1832

Sadly, Mr. Abbott was well-connected, and a fine of £9,000 was reduced to a mere £500. His friend, the Dean of Canterbury, pointed out that this was a matter that affected only ale-drinkers, clearly seeing this as an extenuating circumstance.

It seems almost everything was being colored in some fashion to make it more attractive to customers. Vegetable dyes were often used, but vegetables themselves might be colored with copper salts. Confectionery might include a range of chrome yellow, Prussian blue, copper, and arsenic compounds. Butter was colored with aniline dye, and today you can hear how devious dairy interests prevented margarine makers from coloring their product to look like butter—given the likely dyes, the butter makers probably saved the margarine eaters, even as they poisoned the butter munchers. Yellow was always a problem, particularly in yellow- and orange-colored sweets.

Aniline was also involved in a 1981 case in Spain, which has engaged conspiracy theorists for over 20 years. The official (and quite possibly correct) version is that cheap rapeseed oil containing denaturants was being sold for cooking, that it poisoned some 20,000 people and killed about 300. The figures may have been slightly inflated, as there was compensation money to be had if you claimed your illness or a relative's death was caused by the toxic oil. There is some good circumstantial evidence for blaming the oil, but there are also quite a few oddities that make it possible to argue that the oil had nothing to do with it at all.

Rapeseed oil (canola oil to some) had been allowed into Spain for industrial uses, but it was required by law to be denatured by the addition of aniline to stop it from being used in place of olive oil. Some shady operators refined the oil and sold it from stalls in markets as cooking oil. The authorities claimed that aniline poisoned the consumers. Objectors, however, claim that there is no animal model for aniline poisoning, and the symptoms have not been reliably reproduced.

Medical and toxicological studies require a suspected disease-causing organism or toxin to be present in *all* reported cases, and it must be shown to have caused the disease. With a poison, it is usual to test the suspected toxin on animals, rather than humans, but no animal studies with the dubious rapeseed oil produced the same effects. Until you get that direct evidence, it is usual to keep an open mind, but, as we shall see, this was a most unusual case.

There may have been hereditary differences, or the aniline could have been interacting with something else in the victims' diet—but it might have been something else. Most of the objectors believe the real cause was organophosphorus pesticide residues. It was easier, say the objectors, to blame the operators of small market stalls, leaving vegetable growers in the clear.

A *Guardian* article by Bob Woffinden claimed in 2001 that

more than 1,000 died, and alleged scientific fraud on a grand scale. His case was a circumstantial one, and the logic is flawed in places, but he seems to have identified some interesting discrepancies in the official version—and quite a few unexplained smoking guns. The account below follows Woffinden's outline.

The first diagnosis in the outbreak came when six members of the Garcia family were admitted to two hospitals in Madrid in early May of 1981, all apparently with atypical pneumonia. Dr. Muro y Fernandez-Cavada, the director of the Hospital del Rey, rejected the pneumonia diagnosis, arguing that pneumonia in so many family members did not make sense. Then, as more cases came in, all from the same area, he began to suspect food poisoning. Given the locations of the patients' homes he suspected the cause was some food marketed through alternative routes, not the usual retail outlets.

This was a clever piece of epidemiological reasoning, and Muro and his colleagues came quickly to suspect the unlabeled bottles of cheap oil that they found on sale in the markets. They obtained samples from the homes of victims, and sent them for analysis. It was at this time that administrative paralysis began to set in. Dr. Angel Peralta, at La Paz hospital, suggested publicly that organophosphate poisoning was a likely cause. The next morning, he received a call from the health ministry, during which he was ordered not to say any more about the epidemic, and certainly not to mention organophosphates.

There is a remarkably distinctive smell emitted by fearful bureaucrats. It is acrid, rank, and seems to cling to the clothing and the hair. Acting like a pheromone, it drives senior management to form small defensive herds from which to scream homicidally at middle management that they must not tell junior staff who can fix the problem what is going on because everything, including what has just been reported on the radio, is secret.

Anybody who has been subjected to administrative paralysis would recognize what had happened to Angel Peralta: he had threatened some bureaucrat's promotion chances. So of course he was warned off. When Dr. Muro announced, the very next day, that the outbreaks were market-associated, sterner action was needed and Dr. Muro was relieved of his duties as hospital director, a dead giveaway that administrative paralysis has been overtaken by administrative paranoia. Terrified administrators are, by their cultivated inaction, among the most poisonous objects known.

Almost a month later, the Spanish government declared that the contaminated oils were to blame, yet just the day before this announcement Muro and his colleagues obtained their analysts' results. The oils may have been other than what they claimed to be, but they had different constituents, and therefore could not be the cause of the condition. The real cause was more likely something else, bought in the same markets.

To Woffinden, the real truth seems to be revealed in the pattern of hospital admissions, which peaked at the end of May, ten days before the oil was first officially blamed, and a month before the oil was withdrawn. In addition, the aniline-laced oil had been available for some time before the poisonings started, so what was the trigger? Enter Enrique Martinez de Genique, Secretary of State for Consumer Affairs. Like others, he drew up maps, and realized the same oil had been sold also in Catalonia, yet no cases of toxic oil syndrome had arisen there. In short, the oil could not be to blame, he said. Bad move. The first secret of success in a bureaucracy is not to rock the boat. Señor Martinez was sacked.

A husband-and-wife team of epidemiologists, Javier Martinez Ruiz and Maria Clavera Ortiz, had also mapped the disease distribution and noticed the Catalonian anomaly. They had been appointed to a commission of inquiry, they spoke out, they were sacked. Then, to make sure, the commission of inquiry was closed down.

All of this was circumstantial. What was needed was one case of toxic oil syndrome in a person who had been nowhere near the allegedly toxic oil. Woffinden cites two such cases: one, a woman whose oil came from the olive groves of her Andalusian relatives, the other a lawyer whose husband was certain they had only ever used reputable oils. The lawyer's symptoms were identical to those of the main group of victims, but when it was realized her symptoms had developed 18 months before the main outbreak, she was removed from the list of victims.

Muro and his colleagues went to the markets, talked to stallholders and truck drivers, and concluded that the problems arose from tomatoes grown in Almeria, in the southeast of Spain. This is a dry area but, thanks to plastic tunnels and underground water and chemicals, remarkably productive. If the culprit was indeed organophosphorus poisoning, a reasonable case can be made for this as the source, but Muro died in 1985 and contaminated oil still gets the blame.

I generally prefer not to share the trail with conspiracy theorists and investigative journalists, because they draw bows too long for me, but Woffinden would appear to have outlined a case that is yet to be answered. Either way, members of the public were poisoned by something, but it appears the real poison here was a political fear of being found out, with a synergistic helping of administrative paranoia. If, as Woffinden avers and I suspect, the truth has been hidden, the poisoning could happen again.

There is a delightful case (for students of the matter, not the participants at the time) that was revisited in the *Pharmaceutical*

Journal in 2001. Ian Jones describes the case of Humbug Billy, who accidentally poisoned some children in Bradford, England, in 1858. This character, more formally known as William Hardaker, operated from a stall in the Green Market in central Bradford, and he bought his candies ready-made from one Joseph Neale. They were humbugs, lozenges containing peppermint oil in, theoretically, a base of sugar and gum.

Despite the fact that the price of sugar had been falling steadily throughout the nineteenth century, it was still too high for the cannier confectioners, who eked it out with a filler known as "daft." This was usually plaster of paris, powdered limestone, or gypsum, none of them particularly harmful. In this case, problems arose when a lodger of Neale's, James Archer, made the five-mile trip to buy some daft. In the absence of the druggist, Charles Hodgson, Archer was served by his assistant Goddard, who gave him twelve pounds of arsenious trioxide by mistake.

In those days, there was no control over how arsenic was sold, as long as the sale was recorded, so the poison was held in the same stockroom as innocuous stuff like daft. This is how such errors occurred. Knowing no better, Neale used the arsenic to make up a fresh supply of candies, and when the sweets looked different, presumably because of the arsenic, he sold them at a discount to Humbug Billy.

Humbug Billy himself must have had a few doubts about the odd-looking sweets, as he tried one and was promptly sick, but a bargain is only a bargain if you sell what you buy, so he still sold them. The result was that he made perhaps 200 people severely ill, and killed 20. The death toll could have been much higher, but once the police realized the deaths were not due to cholera—as was first thought—they were able to trace the source fairly quickly. Goddard, Hodgson, and Neale were charged with manslaughter but were later discharged, and Humbug Billy, who should have known better, given his own illness, wasn't even charged. Even when the pharmacists are taking care, poisoning and contamination can happen, but when people cheat, poisoning and death can happen and cover-ups are likely.

The price of sugar again featured in a case in Manchester, England, at the end of 1900, when local beer was found to contain arsenic. Tests later established that glucose had been used in the brewing. Sulfuric acid had been used to convert sucrose, normal sugar, to glucose, but the acid was prepared from iron pyrites rather than pure sulfur and this mineral was the source of the arsenic. The brewers would not have bothered if the price of sucrose hadn't fallen so far that it was now an economical raw material. In short, a worldwide overproduction of sugar led to some 6,000 people being made ill and 70 dying in that single incident.

A consumption of six pints of beer a night was not uncommon, and that would have delivered 45 milligrams (mg) of arsenic. The lethal dose for an adult is typically estimated at about 200 mg, or a week's drinking. After that incident, beer was tested more often, and chemist Sir William Tilden reports that in 1911–12, 1,046 samples of English beer were tested for arsenic. Only 18 of these exceeded the specified level of 1/100 of a grain of arsenious oxide per pound in solids, or per gallon in liquids. In these cases, Tilden believed that the arsenic came from the fuel used to dry the malt in the kiln.

In ancient Rome, soured wine was sweetened with lead acetate, and lead could still be found in drinks as late as the nineteenth century. In 1745, Thomas Cadwalader wrote his *Essay on the West-India Dry-Gripes*, an account of chronic lead poisoning from rum distilled through lead pipes. In 1767, Sir George Baker reported that Devonshire colic was due to the lead lining of the cider apparatus, but the time was not ripe for any centralized control of food, and the wicked grocers could continue to have their way. Taylor mentions deaths from lead contamination of Worcestershire cider as late as 1864.

A little earlier, in 1850, sugar refiners had proposed treating raw sugar with a salt of lead as part of the process of removing impurities from the juice of the crushed sugar cane. This use of lead was known in Egypt around AD 1000. The presence of lead could be detected by exposing suspect sugar near a privy. If the sugar was leaded, the lead salts would react with hydrogen sulfide to produce black lead sulfide (the same chemical reaction that powers lead-based hair dyes). The proportion of lead contained in sugar refined by this process varied from one-twentieth to one-tenth of a grain in a pound. The commissioners of Inland Revenue, considering that even this small quantity might affect the public health, referred the consideration of the question to three men: Pereira, Carpenter, and our old friend, Professor Taylor.

They concluded that even this small degree of contamination, in such a universal article of food as sugar, was objectionable, and that the use of lead should be prohibited as firmly as possible. In any case, it was not so much the quantity of lead taken that determined the symptoms of lead poisoning, said Taylor. Rather, it was its continued ingestion.

Taylor offered up this cheery tale of lead acetate contamination in bread as a salutary lesson:

> . . . about thirty pounds of this substance were mixed at a miller's with eighty sacks of flour, and the whole was made into bread by the bakers and supplied as usual to their customers. It seems that no fewer than 500 persons were attacked with symptoms of poisoning after partaking of this bread. In a few days they complained of a sense of constriction in the throat. . . . The mental faculties were undisturbed. Not one of the cases proved fatal, but among the more aggravated, there was great prostration. . . .
>
> *Lancet*, May 1849

It seems the lead acetate was unevenly distributed in the bread, so we can't be sure how much was consumed, but at least some of those affected must have been lucky. Still, lead was in more foods than people realized. Taylor mentions that pork was sometimes salted in leaden vessels, leaving a residue of lead in the meat. He also mentions that people were still adding litharge (lead monoxide) to sour wine, to give it a sweeter taste. In fact, this must have been fairly common.

Other common sources of lead in the Victorian era were vinegar, which might contain as much as 2 percent lead (along with small amounts of arsenic, copper, and sulfuric acid), and tobacco that had been packed in "patent tin foil," in reality tin-

Once it is absorbed, 97 percent of the lead is taken up by the red blood cells, where it has a half-life in the body of two or three weeks. Some of it ends up in the liver or the kidneys, and some is taken up in hydroxyapatite, a mineral found in bones and teeth. That absorption means we can assess past exposure by X-rays of skeletons and teeth, while urine and blood analysis will reveal current exposures in the living.

We can also assess the lead loads experienced by humans of the past, when we come upon their remains. These days, the normal lead levels for the USA are in the range 0.15 to 0.7 micrograms per milliliter, with an average of 0.3 micrograms per milliliter. The threshold for toxicity is just 0.8 micrograms per liter, but biochemical effects happen at lower levels than this. In the past, many people operated at much higher levels.

coated lead sheeting that would leave a small amount of lead carbonate. Not all of the human lead dose came from food, because the lead industry has always been profitable and poisonous, as we will see in chapter 7. Water, too, would have been contaminated from traveling through lead pipes, but the lead would only have been an additional contaminant.

Clean water sources will become increasingly important, and scarce, throughout the first half of the twenty-first century. In some parts of the world, unchecked population expansion will compel people to use more groundwater, and this may come at a price if more water is being drawn from the aquifer than is added, but there are other traps as well.

In Bangladesh, for example, overpopulation has resulted in grossly polluted surface water, leading to frequent bouts of dysentery and worse. Shallow wells tend to carry the same unpleasant loads as surface water: enter the World Bank and tube well technology. This method has several variations, but each of them results in a small-diameter tube dropping down into clean and unpolluted water. The water was, undoubtedly, clean in the germ-free sense and it was readily available, but it resulted in many Bangladeshis drinking water that well and truly exceeded the World Health Organization (WHO) guidelines for dissolved arsenic.

By the time this became apparent, in the early 1990s, some people had been drinking arsenical water for as long as twenty years. So why hadn't anybody noticed? The short answer is that arsenic builds up in the system and the victims sicken slowly, with none of the sudden symptoms of classical poisoning you would see with a massive dose of rat poison. This buildup of arsenic caused slow, equally lethal damage to selected organs of the body, and a variety of cancers, but the level of medical support in rural Bangladesh was insufficient to detect the pattern.

The problem would have been picked up sooner in the developed world, where all deaths are looked at closely, as are certain sorts of disease. The results are plotted on graphs and examined closely by epidemiologists, to see if any patterns can be detected.

This luxury is not available in the underdeveloped world, and the telltale pattern did not show up. To make matters more difficult, only tube wells drawing water from a depth of 65 to 200 feet carry heavy loads of arsenic and, even then, the dosage can vary markedly. As the water table rises and falls, the dosage may change again. In short, there was no smoking gun, no clear pattern to show up in the records in the far-off capital city, just dispersed illness and slow death from causes apparently unrelated to poison.

The poisoner in this case is nature, though plenty would like to blame the World Bank. The World Bank is a favorite target of activists, because so many of its plans are big plans, said to be unsuited to the developing world. On the subcontinent, the bank is not highly regarded, as Australian poet Mark O'Connor told me in 2003:

> After lunch we drove out through the *Prosopis* thornbush to the Barefoot College, a famous grassroots movement that spreads practical technology in rural India. They told me firmly that they avoid troubling poor people with too much "paper-learning," and they find the World Bank reports extremely useful—for turning into papier maché dolls for village children!

According to activists, tube well water was referred to at first as *devil's water*, which they took to indicate some traditional knowledge of deep water being dangerous. There is no real evidence for this assumption, and it seems equally likely that it was so called only because it came from a considerable depth. The activists also say that when the problem was detected, cover-ups were attempted at first, and this does seem to have been the case.

You can cover up a one-off mistake but not an ongoing poisoning. An estimated 87 percent of Bangladeshis now have access to a tube well within 500 feet of their homes, and there may be as many as 10 or 11 million tube wells. A bad mistake has been made, which now has to be fixed. The mistake was a reasonable one, as the previous water sources were fecally contaminated surface pools, and about a quarter of a million children were dying each year from water-borne disease.

Groundwater had been used in the past, but it was taken from shallow "dug wells" that took only recent rainwater as it sank down, water lying in sediments leached clean of arsenic. Deeper down, the arsenic still has to be flushed out of the 65- to 200-foot sediments. Below 200 feet, the sediments are ancient, flushed, and largely safe. If there is extensive pumping from the lower levels, however, arsenical water will sink down to replace what has been pumped out. Cases of previously "clean" wells showing higher levels of arsenic are now being reported.

The most urgent needs to remedy the problem are: developing some sort of indicator paper that will reliably test water for arsenic content; developing some simple and robust means of clearing most of the arsenic from the water (clearing all of it would be better, but the chemistry means this is unlikely); and instigating a massive drive to sink more deep wells that will draw safe water. There is also a need for consensus on the reason why the water from 65 to 200 feet contains so much arsenic: it is probably a matter of air getting into the soil and releasing the arsenic, but it may also have something to do with fertilizers and organic matter added to the soil.

So where does the arsenic come from? Geologically, Bangladesh is a broad swathe of sediment, and groundwater drifts slowly through it on its way to the sea. Some of the sediment is old, some of it is recent: the older sediment has had most of the arsenic taken out, while any arsenic in surface sediments has long since interacted with air in times of drought, or with organic matter, and turned into soluble arsenic that has leached away.

The piece that is still to be fitted into the puzzle is establishing the actual cause of the release at this time. There is quite a lot of arsenic in the Earth's crust, often found as arsenopyrite in iron pyrite rocks. This is the form in which it is present beneath West Bengal. One widely held theory is that as groundwater is pumped out and the water table is lowered, air infiltrates the sediments, oxidizing the arsenic sulfides and releasing the previously insoluble arsenic in soluble forms. The concentrations are still low, but after an outbreak in Taiwan, the WHO revised its maximum limit for arsenic in drinking water from 50 milligrams per liter to 10 milligrams. To find the Environmental Protection Agency's (EPA) current maximum level for arsenic (and other contaminants) in the United States, go on the Web to www.epa.gov and search under "water."

There are ways in which contaminated water can be treated to bring it within the strict WHO guidelines. For example, carrying and storing water for about three hours in a plastic jerry can that contains a small packet of iron filings will generally remove most of the arsenic, and the packet of filings can be used for 100 days before it has to be replaced. Trials of this method in Nepal gave disappointing results, however, probably because of low sulfate levels in the water. In such cases the addition of small amounts of gypsum (hydrogenated calcium sulfate) may be needed. In laboratory tests, the presence of a sulfate in the water enhanced arsenic removal, while a phosphate suppressed it.

Nature's additives were sometimes augmented and the unpleasant items in your drink put there deliberately. The most common example of this was sailors shanghaied onto ships. Charles Dickens introduces us to such practices as "hocusing":

"Nothing?" said Mr. Pickwick.

"Nothin' at all, Sir," replied his attendant. "The night afore the last day o' the last election here, the opposite party bribed the barmaid at the Town Arms, to hocus the brandy-and-water of fourteen unpolled electors as was a-stoppin' in the house."

"What do you mean by 'hocusing' brandy-and-water?" inquired Mr. Pickwick.

"Puttin' laud'num in it," replied Sam. "Blessed if she didn't send 'em all to sleep till twelve hours arter the election was over. They took one man up to the booth, in a truck, fast asleep, by way of experiment, but it was no go—they wouldn't poll him; so they brought him back, and put him to bed again."

"Strange practices, these," said Mr. Pickwick; half speaking to himself and half addressing Sam.

Charles Dickens, Pickwick Papers, 1837

The practice of using knockout drops, Mickey Finns, or whatever, is by no means a new one, and laudanum was not the only poison available. In India, says Taylor, the seed of the datura was commonly used, though once in a while the poisoner ended up caught in his own trap:

Bassawur Singh, a professional Indian poisoner, ate some of the poisoned food to lull suspicion. In due course, his victims fell insensible, and he robbed them, but after they came around and reported the theft to police, the thief was found about a mile away, quite insensible—and he never came around. All the stolen property was recovered, along with a supply of seeds.

Alfred Taylor, Principles and Practice of Medical Jurisprudence, 3rd edition, 1883

There are two species, he tells us, *Datura stramonium* and *Datura alba*, and he explains that the bitter taste was often hidden by serving the drug in a curry "or in some other highly-flavored article of food." The Thugs of India, he notes, commonly use the seeds of *D. alba*.

Another drug used in England for hocusing was a decoction of the levant nut, *Cocculus indicus*. This was also sold mixed with grain as "Barber's poisoned wheat," but there was another cunning use of the poison, extracted from the plant's berries. Porter, ale, and beer sometimes owed their intoxicating properties to this extract, a practice of which Professor Taylor clearly did not approve:

> The fraud is perpetrated by a low class of publicans. They reduce the strength of the beer by water and salt, then give to it an intoxicating property by means of this poisonous extract. A medical man consulted the author some years since in reference to the similarity of cerebral symptoms suffered by several of his patients in a district in London. It was ascertained that they were supplied with porter by retail from the same [public] house.

Alfred Taylor, *Principles and Practice of Medical Jurisprudence*, 3rd edition, 1883

In short, some people were getting a different poison, picrotoxin, rather than ethanol in their beer, producing symptoms similar to those of drunkenness, but far more cheaply, and free of any excise.

Those poisoned want to sleep, yet at the same time they are beset by wakefulness and a heavy lethargic stupor. They retain a consciousness of passing events, but feel a complete loss of voluntary power, according to Taylor. Slipping foreign material into food was something of an art in the nineteenth century, but almost everywhere you turned then, you were beset by poisons.

Anamirta cocculus (paniculata)

Strange foods have always posed a problem for travelers, where there was no precedent, no standard way of dealing with them. When he was visiting Australia in 1770,

Joseph Banks noted how the crew had been so long at sea "with but a scanty supply of fresh provisions that we had long usd to eat every thing we could lay our hands upon, fish, flesh or vegetable which only was not poisonous." Later, he recorded his experiences with seeds of a cycad, which he believed would be safe to eat, going on the evidence of them being eaten regularly by the locals:

> By the hulls of these which we found plentifully near the Indian fires we were assurd that these people eat them, and some of our gentlemen tried to do the same, but were deterrd from a second experiment by a hearty fit of vomiting and purging which was the consequence of the first. The hogs however who were still shorter of provision than we were eat them heartily and we concluded their constitutions [were] stronger than ours, till after about a week they were all taken extreemly ill of indigestions; two died and the rest were savd with difficulty.
>
> Joseph Banks, *Journal*, 1770

Banks was unaware of the proper methods for preparing the seeds. He learned better when he called at Prince's Island (Pulau Panaitan), on the way to Sumatra:

> Their food was nearly the same as the Batavian Indians, adding only to it the nuts of the Palm calld *Cycas circinalis* with which on the Coast of New Holland some of our people were made ill and some of our hogs Poisond outright. Their method of preparing them to get out their deleterious qualities they told me were first to cut the nuts into thin slices and dry them in the sun, then to steep them in fresh water for three months, afterwards pressing the water from them and drying them in the sun once more; they however were so far from being a delicious food that they never usd them but in times of scarcity when they mixt the preparation with their rice.
>
> Joseph Banks, *Journal*, 1770

Rule number 1, then, is to assume that the food may need some preparation; rule number 2, to assume the locals know what to do; and rule number 3, to make sure you find out what they do. Ludwig Leichhardt, the Australian explorer, described the preparation of *Cycas* in these terms:

> I also observed that seeds of *Cycas* were cut into very thin slices, about the size of a shilling, and these were spread out carefully on the ground to dry, after which (as I saw in another camp a few days later), it seemed that the dry slices are put for several days in water, and, after a good soaking, are closely tied up in tea-tree bark to undergo a peculiar process of fermentation.

> Ludwig Leichhardt, *Journal of an Overland Expedition in Australia*, 1846

Leichhardt would try foods found in the crops of parrots he shot, assuming that what was safe for them might be safe for him. With some birds, this could be a little risky, given that cassowaries in northern Australia eat the blue fruits of the cassowary plum, *Cerbera floribunda*, with impunity because the poisonous seeds pass through with no harm to seed or bird. An examination of a dead cassowary's crop might easily suggest, falsely, that the seeds were harmless.

Around 2,000 plant species contain the cyanogenic glycosides found in *Cycas*, so when the plant cells are crushed, wilted, or frozen, enzymes are released that hydrolyze the glycosides to cyanide. Examples include sorghum, corn, some of the clovers, and, in particular, cassava. The glycoside in cassava is called linamarin. If untreated cassava is eaten, this will cause a disease known in rural Nigeria as *konzo*. Over time, people there have learned to leave cassava to soak for some days to allow the linamarin to be broken down, either by enzymes in the cassava or by lactobacteria.

Konzo causes irreversible paralysis of the legs. The name means "tired legs," because of the ways the knees are drawn together. Typically a person with konzo walks leaning over backward, but some of the worst-affected children can only crawl. One of the challenges facing agribiologists at the moment is to breed a strain of cassava with reduced amounts of linamarin while still retaining the disease resistance, high yield, and palatability of the original forms.

The science of the nineteenth century concentrated on identifying and detecting poisons; that of the twentieth century on developing antidotes to and syntheses of organic poison. The science of the future will be aimed at decreasing the natural and synthetic poisons in our foods and environments.

Before we look at poison as medicine, with its accompanying risk of "kill or cure," it would perhaps be advantageous to investigate the science of poison through the ages.

4

THE SCIENCE
OF POISON

One day [Caligula] sent a colonel to kill young Tiberius without
warning; on the pretext that Tiberius had insulted him by taking an
antidote against poison—his breath smelled of it.

Suetonius, *The Twelve Caesars, c.* 110

In any place where poisoning is common or believed to be
common, there will be a strong motivation to study poison.
Where some work in secret to discover poisons, others will seek
and claim to find ways of counteracting them. So Rome had a
flourishing trade in poison cures, some more effective than
others. When Cleopatra poisoned herself by clasping an asp to
her breast, Octavian, better known to us by his later imperial
name, Augustus, felt deprived of the centerpiece for his tri-
umphal procession back in Rome and called in Psyllian snake-
charmers to suck the poison from her wound. But there would be
no shameful display for the Queen of the Nile, who was reputed
to be no mean poisoner herself.

Accidental poisoning was a major problem in the search to
detect and identify individual poisons. Alfred Taylor described

three cases of poisoning from nitrate of aconitine. The first victim was given this preparation as a treatment for chronic bronchitis, the second (who survived) was given a prescription by the same doctor, and the third was the doctor in the case, a Dr. Mayer, who also took what should have been a safe dose and died. It turned out later that the dispenser had used a supply from Paris, rather than the regular German solution known as Friedländer's nitrate of aconitine. The Parisian tincture was 170 times as strong, something that could only be shown by systematic tests and experimentation.

Poisons have always held just as much fascination for scientists as they have for the lay public. There seems to be a special sense of achievement to be gained from overcoming the secret power of poison, perhaps even more than in becoming a master of the application of the power. Scientific advances have led to a common assumption today that, given a poisoning, medical staff need merely to reach for the appropriate antidote and apply it, and the case is solved. The more unpalatable truth is that most poisons lack an antidote, although there are treatments available for some sorts of poison.

Evolution plays many excellent tricks with poisons, mainly because there are so many easy targets in the biosphere, and almost every poison has a unique way of wreaking havoc on our cells.

A living cell is far more than the docile bag of uniformly dilute soup with a nucleus in the middle we see in most textbooks. Each cell is a teeming commonwealth of chemical bits and pieces, packed with molecular citizens, some of them bustling porters hauling needed material from A to B, some of them cunning artisans, accepting the porters' raw materials and fashioning complex molecules from the bits according to simple plans, laid out in yet other molecules that are passed to them. The molecules that make up the cell dance around each other, signal, and interact with each other and cooperate in a myriad ways.

Even the membrane sealing off one cellular city-state from its surroundings is far more than a simple bag, because on the cell's perimeter molecular guardians and gatekeepers watch their surroundings. They challenge some approaching molecules to prove their worthiness to enter, while others are recognized and dragged roughly inside, because there is a need for them. Some chemical guards communicate with the interior to find out what is needed, others either allow appropriate molecular messengers free passage out of the cell, or speed unwanted chemicals on their way.

The cell is a coalition of chemicals working as one to maintain what we call "life" within the cell. Most cells are part of a larger nation, and each cellular city-state stands ready at any stage to pull the plug on itself by releasing the poisons it holds ready. And what is the trigger for this release? Usually, it comes from chemicals, inside or outside the cell, rogue molecules that mug the citizens. Some enemies block the gateways, others force their way in and attack the artisans and the porters. In short, the triggers are the things we call poisons. A healthy cell lives a poison-free life, but when there is a need, it will end its life in a withering fire and counter-fire of poison.

There exist only a few genuine antidotes, chemicals that take a stronger grip on the poison than the body can, either by locking it up or keeping it away from our cells, in much the way that the minor poison ethanol is a treatment for diethylene glycol. Atropine is good for the carbamate and organophosphate insecticides, vitamin K is good for anticoagulants such as warfarin rat poison, while acetic acid counteracts "1080" or sodium fluoroacetate. Arsenic and a few other compounds are limited by British Anti-Lewisite. Thiosulfate and nitrite (itself a poison) are antidotes for cyanide, while methylene blue counteracts the poison effects of nitrite, but that is about the extent of it. Antidotes are rare indeed.

Acids and alkalis may be neutralized, but most other treatments are designed either to purge the body by clearing the poison from the alimentary canal, or, occasionally, to absorb it. Boardinghouse keepers in colonial Australia were in the habit of adding charcoal to the salt meat if they suspected it was "off," in the hope the charcoal would take up the smell. If their poor residents became ill, the doctor at the hospital would probably dose them with charcoal "biscuits" in the hope that these would absorb the poison. Burnt toast, in effect activated charcoal, is still used in some poison cases. That aside, the dream of antidotes and universal antidotes is mainly mythology, a myth A. E. Housman tells best:

There was a king reigned in the East:
There, when kings will sit to feast,
They get their fill before they think
With poisoned meat and poisoned drink.
He gathered all the springs to birth
From the many-venomed earth;
First a little, thence to more,
He sampled all her killing store;
And easy, smiling, seasoned sound,
Sate the king when healths went round.
They put arsenic in his meat
And stared aghast to watch him eat;
They poured strychnine in his cup

And shook to see him drink it up:
They shook, they stared as white's their shirt:
Them it was their poison hurt.
—I tell the tale that I heard told.
Mithridates, he died old.

A. E. Housman, "A Shropshire Lad, LXII," 1896

The real Mithridates was king of Pontus in Asia Minor, the place where the bees made poison honey (see chapter 10). He lived from 114 to 63 BC, and, even in death, his story held a special fascination for the Romans, especially Pliny. It seems he tried out various poisons on condemned criminals, and he also tested antidotes to the poisons by administering the alleged antidotes either before or immediately after the victims were given the poison. Just as Housman tells us, he also took small doses of poisons each day to make himself immune.

The formula for his treatment was known as "Mithridatum," and he guarded his secret jealously until Pompey invaded Pontus and took it back to Rome. Pliny's recipe for Mithridatum lists some 54 poisons, including one he described as "the blood of a duck found in a certain district of Pontus, which was supposed to live on poisonous food, and the blood of this duck was afterwards used in the preparation of the Mithridatum, because it fed on poisonous plants and suffered no harm."

The high point of the Mithridates legend for the Romans was that during Pompey's invasion Mithridates attempted to commit suicide by poison. The poison failed, perhaps due to this antidote, so they say he made his bodyguard stab him to death.

Mithridates was following in the scientific footsteps of Nicander of Colophon, who lived somewhere around 185–135 BC and was physician to Attalus, King of Bithynia. Nicander also experimented with poisons on condemned criminals. From these experiments, Nicander wrote two volumes, *Theriaca* and *Alexipharmica*, on antidotes to poisonous reptiles and poisons respectively. Ceruse, litharge, aconite, cantharides, colchicum, hemlock, henbane, and opium were among the 22 he listed. Oddly, Nicander's *theriac* (literally, something to do with wild beasts and venomous reptiles) is a word we still use today, though its use has mutated and it no longer means snake repeller. Beginning with straightforward theriacs like the theriac of Andromachus, over time different cities developed their own special formulations, not that theriacs were needed everywhere. The Venerable Bede lived after Saint Patrick and mentions Ireland's lack of snakes without invoking Ireland's patron as the cause:

> No reptiles are found there, and no snake can live there; for, though often carried thither out of Britain, as soon as the ship comes near the shore, and the scent of the air reaches them, they die. On the contrary, almost all things in the island are good against poison. In short, we have known that when some persons have been bitten by serpents, the scrapings of leaves of books that were brought out of Ireland, being put into water, and given them to drink, have immediately expelled the spreading poison, and assuaged the swelling.
>
> Bede, *Ecclesiastical History*, c. 731

By 80 BC, one Zopyros offered a theriac named "ambrosia." It was said to include frankincense, galbanum, pepper, and other aromatic substances, all incorporated into boiled honey as a sort of lozenge and to be swallowed with wine. After that, theriacs seem generally to have been sweet and thick, and by 1440 the adjective "theriacal" was in use. The *London Pharmacopoeia* of 1746 has a formula for a theriac first made in Italy and now called Venice treacle, with over 70 ingredients, including dried vipers, roses, licorice, spikenard, myrrh, horehound, pepper, valerian, gentian, St. John's wort, germander, galbanum, and 75 percent honey.

By the end of the eighteenth century, theriacs were just all-purpose medicines, given for all of the poisons causing disease—until the germ theory was established, *every* disease was, by definition, caused by a poison of some sort and had to be treated with an antidote and lots of purging.

If doctors and scientists could not find antidotes or cures for criminally inflicted poisons, they might prevent some deaths just by showing they could detect poisons, even in tiny doses. This set up a sort of arms race, with the poisoners seeking out new poisons with as little taste as possible, or a taste that could be disguised, and the doctors and scientists seeking ways of recognizing trace elements. Today, we have mass spectrometers, but the chemistry of the early nineteenth century had to develop a test that would prove the presence of a given chemical in a residue.

Most of the tests destroyed the remnants that were being tested, so there was usually just one chance of getting it right. The test needed to show a reaction to one element or one chemical only, and it should not show any reaction in the absence of this element, which meant using very pure chemicals in the analysis. Aside from that, in the nineteenth century the best people could do was to set down carefully the effects of poisons. These were scientists of great dedication, so we should not be too surprised if they sometimes used themselves as guinea pigs.

In 1867, for example, John Harley reported on some of his observations after poisoning himself slightly with hemlock. He generally confirmed the description of rising paralysis given in Plato's description of the death of Socrates, a matter we will come back to in chapter 8. Harley also indicated that he considered hemlock to be a suitable treatment for the children we would now call hyperactive—a somewhat surprising alternative to Ritalin!

Others also were willing to risk their own health to save lives. The disease pellagra was first reported in Spain in 1735, and while we know now that it is a deficiency disease, it took people a long time to discover this. Pellagra seemed to be associated with eating a diet high in corn, the marvelous crop that had been brought to Europe from the Americas and that had become the staple of the poor of Europe. Perhaps the corn was diseased, poisoned in some way, but nobody got any closer to a cure while working on this theory.

In 1914, the American scientist Joseph Goldberger noted that only poor people got the disease. He then detected a strange pattern of pellagra victims in an orphanage: only children between six and twelve years old got the disease. It turned out that those under six were given plenty of milk, and those over twelve were given more meat, but those between six and twelve were largely eating corn, corn, and more corn.

Everything pointed to some poison or dietary deficiency, but most people "knew" by 1914 that diseases were caused by germs, so this must be just a fussy germ. Yet if there really was a germ causing pellagra, why did it only attack children in a particular age group? Given the task, how would *you* prove there is no pellagra germ?

Goldberger knew how: he assembled a team of sixteen volunteers, including himself and his wife, and they tried to infect themselves with pellagra in a variety of ways. First, they tried blood transfusions from pellagra sufferers, then they took swabs from the sufferers' noses and throats and applied these to their own noses and throats.

Last of all, the volunteers swallowed pellets of dough made from the victims' urine, feces, and scrapings from the pellagra sores, mixed with some flour. Nobody developed pellagra, and people generally agreed the researchers had proved pellagra was a deficiency disease. History does not tell us whether any of Goldberger's opponents dared to repeat his drastic experiment to try to prove him wrong.

We have strayed, though, from the question of poison detection. It was a popular topic, and even the man known best today as composer Alexander Porphyrevich Borodin, in everyday life a professor of chemistry, wrote his doctoral dissertation on the similarities between arsenical and phosphoric acids. This, however, was of limited use in proving poisoning in a court of law, but English chemist James Marsh's research was more helpful.

In 1832, one John Bodle was accused of putting arsenic in his grandfather's coffee, and James Marsh was called to perform the then standard test of bubbling hydrogen sulfide through the suspect solution. The precipitate he obtained did not keep well, and the jury was not convinced: Bodle went free, although he later confessed to poisoning his grandfather. Marsh was so angry Bodle had escaped that he devoted himself to developing the reaction we now call the Marsh test.

Four years later, Marsh was working at the Royal Arsenal at Woolwich, and it was there that he developed his accurate test for revealing traces of arsenic in the body. It was exquisitely sensitive and could reveal as small an amount as 1/50 mg in a sample taken from the body. Alfred Taylor offers this technical description: "The action of this test depends on the decomposition of the soluble compounds of arsenic by nascent hydrogen, evolved from the action of diluted sulphuric or hydrochloric acid on zinc."

Translated, Taylor's explanation was that dilute acid, acting on zinc, will produce hydrogen. The hydrogen will reduce arsenic compounds to a small amount of elemental arsenic that will then be deposited on glass after heating. This deposit would be the proof positive that arsenic had been in the sample tested. For chemists, the reaction involves producing arsine gas, which then diffuses along a heated tube, where it decomposes to leave a silvery-black film of telltale metallic arsenic.

The art of testing had come a long way in a scant hundred years. When Mary Sherman was tried at the Old Bailey in 1726 for the murder of Mark Shovat with "a great Quantity of white Arsnick, being a deadly Poison, with Milk and Water, and sending the same to him to drink," a person referred to as Mr. Belcher the Apothecary was called. He found no sign of poisoning in the autopsy:

> I saw not the least Symptom of Poison, and verily believe that he dy'd a natural Death; I tasted some of the Emulsion that was left, it was grown a little Acid with keeping. I could not discover, that there was any thing Poisonous, or Hurtful, but I took it to be a Cooling Innocent Draught, made of Almonds, Barly-water, Orange Flower-water, Sugar, and some other harmless Ingredients.—I advis'd them to make some tryal of its Effect, upon a Puppy, or a Kitling, but they said, there was not enough left, to make the Experiment.—I cannot think that it was the Cause of the Deceaseds Death.

> Mr. Belcher the Apothecary, 1726 (OldBaileyOnline.org)

While dogs, ducks, and other animals were often used to test for poisons, by the nineteenth century we can see glimmerings of the antecedents of the rudiments of systematic chemistry. This novel science allowed new tests to be contemplated or even applied, and a few of the old methods could be discarded, but change was, and is, a slow process.

As late as the 1880s, the standard test for cantharides or Spanish fly began with an aqueous or alcoholic extract from the stomach contents. This extract was heated gently to reduce it, then placed on the experimenter's arm, earlobe, or lip and

covered with goldbeater's skin, a collagen membrane traditionally used to mend vellum manuscripts. If cantharides was in the
extract, when the place was uncovered after three or four hours
and wiped with chloroform a blister would have formed.

One of the most effective tests developed in the nineteenth
century is the flame test, which was discovered by Robert
Bunsen, working with Gustav Kirchhoff. Together, they created
the science of spectroscopy, examining the colors produced when
samples of different elements are heated in the flame of what we
now call the Bunsen burner (although it was probably invented
by Bunsen's assistant, Peter Desaga). Today, we use similar
methods to find out about the chemicals in space, far beyond the
reach of our laboratory tools. (By the way, Bunsen lost an eye in
an explosion while investigating cacodyl cyanide, a poison that
was later proposed for use against the Russians during the
Crimean War.)

Each metallic element produces a characteristic color, and
there is a probably apocryphal tale that Bunsen suspected his
landlady of recycling one night's scraps in the next night's
stew—a sure way of causing food poisoning. So he sprinkled
some leftover meat scraps on his plate with lithium chloride.
The next night, a sample of the stew was found to give the brilliant red flame test characteristic of lithium. Lithium, as Bunsen
pointed out, is not characteristic of stews. Well, so runs the
legend—and it probably happened to somebody, once or twice.

The flame test can detect other metallic poisons, and, once, it
even detected a poison that did not yet exist when William
Crookes was working with selenium ores. After he heated a
sample, it showed in its spectrum a bright green line characteristic of no known element. Crookes reasoned that it had to be a
new, hitherto undiscovered element, and he reasoned correctly,
because there were traces of thallium present in the selenium ore.

He named it thallium (from the Greek word *thallos* meaning "green twig") after the color of its line. The botanical word *thallophyte* comes from the same root.

Crookes felt entitled to claim the discovery of thallium, but a Frenchman, Lamy, was the first to isolate the actual metal and so gets the credit for discovering thallium. Lamy reported later that he had dissolved 77 grains of thallium sulfate in milk, and he found this quantity was sufficient to kill two hens, six ducks, two puppies, and a middle-sized bitch.

Crookes, on the other hand, questioned whether his beautiful green thallium was poisonous at all. He reported that even though he had been much exposed to the fumes, they had produced no particular effects. Taylor reports Crookes as saying he had also swallowed a grain or two of the salts without injury.

Thallium has a local action on the hair and skin, staining the former and rendering the latter yellow and horny, say the references, but Crookes reported no such effects. One thing is certain: Crookes's lasting fame came not from thallium, not from his excellent work on the cattle plague, not from his work in pioneering the publication of science news, but from a small toy, still sold today. The Crookes Radiometer is a sort of windmill in a globe, with four vanes, white on one side and black on the other. It is based on an effect Crookes noted while investigating physical effects in a high vacuum. He wanted to weigh very small samples in a vacuum in order to get the atomic weight of thallium and found unexpected movement, caused by gas molecules bouncing off warmer surfaces. The standard explanation of light photons bouncing off the white surfaces can be dismissed by looking to see which way the vanes turn.

Spectroscopy, though, did have its uses in detecting poisons. In 1872, a researcher named Barrett showed how the flame of pure hydrogen is rendered a vivid green by phosphorus, but as most body parts have phosphates in them, this was of little use as a test. Still, in 1972, a Graham Young was convicted of two thallium murders in England when the element was detected in one victim's ashes after cremation. These days, though, spectroscopy is far more sophisticated than it was in Bunsen's time.

Harold Shipman may have murdered 200 patients in Manchester, England, though the true number may never be known. We can, however, be more sure of his guilt, thanks to spectroscopy. When he arrived at Mrs. Kathleen Grundy's home one summer morning in 1998, he pretended to take a blood sample but in fact injected Mrs. Grundy with a lethal dose of morphine.

Checking the records, police learned that Shipman had been going through large amounts of morphine, and this drug was found in many of the alleged victims, but any defense would almost certainly claim these women were all habitual drug users. Enter Hans Sachs, a master chemist and toxicologist of Munich, who applied spectroscopy to hairs taken from the victims. Hair grows about one centimeter a month, and, using a mass spectrometer, Sachs found levels and locations of residues consistent with morphine use on only one or two occasions. Narcotic abusers typically have 2 nanograms of the substance per milligram of hair, but the victims showed levels a mere half of one percent of that, 1/200 of the amount found in habitual users.

Detection of poisons in the future, especially those in gaseous form, will be through devices using a surface acoustic wave detector. This is a small, vibrating crystal that changes its frequency when a minute amount of a chemical agent is absorbed onto the surface. It should have fewer false positives than other devices such as ion mobility spectrometers, which may be fooled by interferants like aftershave, perfume, paint, and other solvents. The life of the criminal poisoner gets ever more difficult.

While a false alarm caused by interferants may not cause a

great deal of harm, a false positive in a trial for murder can be another thing altogether. In the case of Thomas Smethurst in 1859, Alfred Taylor discovered he had made an error. He had used impure copper in the Marsh test, so he himself had introduced the arsenic he detected in the test. Smethurst was still convicted, although reprieved as a result of a public outcry, and later Taylor made this very clear point, in the context of the related Reinsch test:

> The copper must first prove to be free of arsenic, as this is a very common contamination of commercial copper in the form of foil, gauze, or wire. Copper-gauze and wire generally contain arsenic. Pure electrolytic copper, free from arsenic, can be procured in the form of thin sheet or foil. If the arsenic is present in the liquid, even in small quantity, the polished copper acquires, either immediately or within a few minutes, an iron-grey metallic coating from the deposit of this metal.
>
> Alfred Taylor, *Principles and Practice of Medical Jurisprudence*, 3rd edition, 1883

Quite often, the toxicologist's target is a poison that harmed or killed an animal, rather than a human, but few hunts could have been as desperate as the chase in 2001 to find what had killed 500 thoroughbred foals in Kentucky. Some of the dead foals were stillborn, others were born alive and died soon after, but it was the same culprit each time: cyanide from wild black-cherry trees, passed on to the foals' dams via caterpillars.

The toxicologists began by examining the horses' feed, then they looked for fungal toxins on grasses, before lighting on Eastern tent caterpillars, which were present in the area in large numbers. Somebody noticed that there were always cherry trees in areas where the deaths were happening, and this proved to be the key to the puzzle. An unusually hot early spring had concentrated cyanide in the cherry leaves, but Eastern tent caterpillars are cyanide-resistant. Having stripped the trees bare, they moved on to the pastures, while they and their feces were still full of cyanide. One way and another, the pregnant mares picked up enough poison to do the damage, researchers concluded, but the cream of the year's racehorses had been lost by the time the deaths were accounted for.

Poison crops up in sports quite often as a quasi-medical performance enhancer, a tradition that may owe its origins to the Styrian arsenic eaters whom we will meet later. Of course, if you regard the Viking pastime of fighting and killing as a sport, then using poisons in sports goes back to the berserkers, who allegedly dined heavily on hallucinogenic mushrooms and then ran, screaming and hewing, into the battle. We can, however, assume that most of the tales of English cyclist Arthur Linton dying in or after the 1886 Bordeaux to Paris race as a result of taking performance-enhancing drugs are wrong: the first race did not happen until 1891, and Linton won the race in 1896, a decade after his alleged death, though he did in fact die several months after the 1896 win, apparently of typhoid fever. Nonetheless, the Linton case is commonly cited in the press as an early example of drugs being used in sports. To this day, we do not know for certain that Linton used drugs in 1896, but people who claimed to know the truth believed that strychnine, heroin, and "trimethyl" were involved in some combination.

Those unscrupulous enough to use drugs in sports are not above using them on the animals they train. Popular legend in the antipodes claims that when the champion racehorse Phar Lap died in the U.S., he was "poisoned by the Yanks" because he was too good, but there is another theory, spoken of occasionally by colorful ancient racing identities in Australia, to the effect that Phar Lap was accidentally poisoned by his own connections, who had been dosing him regularly with strychnine (or sometimes strychnine and another poison—urban legends are like that) to make him go faster. The punch line of this Australian version is that a veterinary surgeon was called in and legitimately gave the great horse a small dose of strychnine. This, combined with the illicit doses that had been building up in his system, was enough to kill him.

Strychnos nux-vomica: *a,* flowering branch (¼ natural size; *b,* cross section of fruit; *c,* corolla; also anther, pollen, pistil, ovary, seed, enlarged.

Strychnine first appears in the annals of athletics during the 1904 Olympics, when U.S. marathon runner Thomas Hicks took a restorative dose of strychnine, egg, and brandy. He collapsed on the track and was lucky to survive. More recently, an Indian woman weightlifter, Kunjarani Devi, tested positive for strychnine in South Korea in 2001 and was banned from competition for six months. One of her associates claimed that strychnine was hard to obtain in India (which happens to be a major importer of *Strychnos nux-vomica* material from Indonesia, from which strychnine is extracted!).

This still leaves boxers, who are said to have used any number of atropine, caffeine, camphor, cocaine, digitalis, heroin, and even nitroglycerin in the past to block pain. Perhaps those old pugs weren't punch-drunk after all, just in training? Most of the use of dangerous drugs and poisons seems to have come in with professionalism: if there is a profit to be made, poison is likely to be on the scene somewhere.

That being said, almost all of the poisons listed above would have been included, quite legitimately, in the medicine chests of the past, and the not too distant past at that.

5

POISON IN THE MEDICINE CHEST

I sent for Ratcliffe; was so ill,
That other doctors gave me over:
He felt my pulse, prescribed his pill,
And I was likely to recover.
But, when the wit began to wheeze,
And wine had warmed the politician,
Cured yesterday of my disease,
I died last night of my physician.

Matthew Prior, *The Remedy Worse Than the Disease*, c. 1715

Poison has traditionally been part of the physician's armory, and the best poisons are surely, for most of us, those that have benefited humanity. The Squibb Pannier was a medical chest designed to be carried into the field by Union medics in the Civil War, and it reflects medical thinking in the 1860s, a time when medical knowledge was changing fast. Nonetheless, practitioners still relied on strong poisons for a strong enemy, because in war at that time, more would die of disease than would die of bullets.

More than half of the drugs in the Squibb Pannier are imme-diately identifiable as poisons: Spanish fly, silver nitrate, iodine, tartar emetic, calomel (mercury chloride), strong alcohol, chloro-form, ammonia, laudanum, quinine, camphorated tincture of opium, iron sulfate, compound cathartic pills, colocynth and ipecac pills, powder of ipecac and opium, potassium chlorate, morphine sulfate solution, camphor and opium pills, blue mass (mercury) pills, creosote, fluid extract of aconite root, fluid extract of colchicum seed, fluid extract of ipecac, ferric chloride, lead acetate, and zinc sulfate were all there, sometimes in more than one form.

If poisons were powerful, they usually had the advantage of tasting bad. The rule has long been that the more evil-tasting a preparation and the nastier its effects, the more efficacious it must be as a cure. Sometimes, of course, the evil-tasting substances can in fact do some good but still taste seriously evil. That is when honey and sugar come into their own, as a means of getting vile-tasting material past the taste buds. (Tea, coffee, and sugar were also in the pannier.)

Cephaëlis ipecacuanha

A common poison that wasn't in the pannier but that is still used today is digitalis. Some believe Vincent van Gogh to have been poisoned by digitalis, *Digitalis purpurea* or the purple fox-glove. Not long before his death, he painted portraits of his homeopath, Dr. Gachet, and his daughter. He later painted a second portrait of the doctor. In both, Gachet was represented with a foxglove, the significance of which is not clear.

There is no evidence that Gachet ever prescribed digitalis for his companion, but some claim van Gogh's penchant for yellow and his instability both stemmed from digitalis poisoning. According to all accounts, however, van Gogh was also fond of eating his oil paints and drinking turpentine and had long been highly unstable and self-destructive. Gachet himself believed van Gogh's unpredictable behavior could be attributed to turpentine poisoning and overexposure to the intense Provençal sun.

Digitalis purpurea

Then again, there is another faction who would rather blame van Gogh's whole situation on absinthe. Absinthe is alcohol adulterated by oil of wormwood and an outlawed beverage in many countries since just before World War I. It can cause forgetfulness, delirium, convulsions, and brain damage, or so we are told. It is emerald green and a favorite with intellectuals and artists, but the thujone it contains was supposedly a serious problem. In fact, it seems more likely the problem was one of bad publicity, given that renowned wicked men such as Charles Baudelaire, Arthur Rimbaud, Aleister Crowley, Paul Verlaine, and Oscar Wilde all drank it! (The modern version, sold as "absinth," is said to lack the hallucinogenic components.)

Absinthe starts with ordinary wormwood, *Artemisia absinthium*, which is soaked in ethanol to extract the thujone and also some horridly bitter substances called absinthins. A distillation of the steep liquor got rid of those, then other herbs (including the Roman wormwood, *Artemisia pontica*) were added, along with the vital green coloring, usually chlorophyll.

The drink's evil reputation was not aided by the infamous van Gogh ear-slicing incident, also blamed, possibly unfairly, on absinthe, or by the alleged but erroneous chemical structure of alpha-thujone, said to be similar to THC, the active ingredient in cannabis. Some now doubt the toxicity of alpha-thujone, because it is only a monoterpene, similar to eucalyptol and menthol. It is a harmless enough substance found in herbs like sage, tansy, and tarragon and even in Vicks VapoRub, a popular chest ointment sold around the world to ease the breathing of young babies. Wormwood itself has long been known to nursing mothers: here is Juliet's (wet) nurse, describing how she weaned her charge.

> But, as I said,
> When it did taste the wormwood on the nipple
> Of my dug and felt it bitter, pretty fool,
> To see it tetchy and fall out with the dug!
>
> William Shakespeare, *Romeo and Juliet*, Act I, scene iii

On the other hand, there can be no doubt that oil of wormwood in large doses was far from benign. Alfred Taylor describes a case where a druggist's shopman was found early one morning by his master, lying on the floor of the shop, completely unconscious, convulsed, and foaming at the mouth. In a short time, he was no longer violently convulsed but remained insensible, with clenched jaws and dilated pupils.

There was a bottle of oil of wormwood nearby, and it was clear he had consumed some of it, perhaps half an ounce (15 milliliters). His pulse was weak, compressible, and slow. On recovering, he had totally forgotten all the circumstances connected with the case and maintained he knew no reason why he should have taken the oil. Taylor suggests he may have imagined himself suffering from worms and sought relief in an unusual dose of this oil.

Assorted species of *Artemisia*, the mugworts, make powerful antimalarials, not for our benefit but to control the roundworms that would otherwise infest their roots. In China, artemisia, in the form of *qinghao*, has been used on malaria for more than a century, and in 1972 Chinese scientists isolated the active ingredient (called artemisinin) from *Artemisia annua*, or sweet wormwood. Since then, quinine resistance problems in Southeast Asia have grown worse, and medical workers have been using artemisinin on a large scale to treat infected humans.

Nobody really knows how it works, but researchers think the artemisinin may kill the *Plasmodium* parasite by producing free radicals when it binds to heme. These highly active chemical fragments may act as a poison and kill the parasite by attacking the proteins and lipids in its cell membrane. Work on new antimalarials is well-timed, as quinine and its derivatives are losing the fight, and that other powerful antimalarial poison, DDT, is under fire from well-meaning reformers.

Until recently, its ready availability made quinine the preferred suicide agent in Greece, and when it was sold over the counter in Britain, it was ostensibly intended for the treatment of colds and fevers but was in fact mainly used to terminate pregnancies. To have any effect on the uterus, however, it had to be delivered at close to toxic levels, and so it was banned in Britain.

Quinine was included in many antimalarial tonics because it tasted so bad, but it actually poisons the malarial parasites only a little more than it poisons us. It is toxic in medium doses, and repeated dosing can lead to cinchonism, which presents as tinnitus, headaches, nausea, abdominal pain, and skin rashes. In a severe malarial case, quinine used to be given at close to a lethal dose and could cause blackwater fever.

Quinine had one unexpectedly deadly effect on humans, in that it allowed Europeans to survive in their equatorial colonies, to the detriment of the locals. Its absence could be equally deadly, as could religious intolerance. In Oliver Cromwell's time, quinine was known as "the Jesuits' bark." For this reason, Cromwell, when struck down with a dose of malaria (common in Britain in those days), declined the Jesuitical treatment and died of ague, ignorance, and bigotry in roughly equal proportions.

Ignorance also seems to have played a role in most or all cases of poisoning with diethylene glycol, or DEG. In 1937, something like a hundred people died in the United States when DEG was used to dilute sulfanilamide. This was before the first of the antibiotics, and sulfanilamide had only just become available as a weapon against previously deadly bacterial infections. The snag was that it was served up as a mixture containing 10 percent sulfanilamide and 72 percent DEG.

Cinchona succirubra

We know now that DEG, most commonly used as an antifreeze and as an industrial solvent, is metabolized by alcohol dehydrogenase (ADH), the substance that protects us from alcohol, into deadly oxalic acid. The acid is fairly mild as acids go, but it forms crystals in the brain, causing irreversible damage, and in the kidney tubules, so victims die of acute kidney failure. Around 2,000 plants contain oxalic acid, including many of the lilies and the oxalis that gives it its name.

The involvement of ADH is an oddity. Because of ADH's role in turning DEG into a poison, this form of intoxication is sometimes *treated* with alcohol, either orally or intravenously. The ADH attacks the ethanol preferentially, leaving the DEG intact and thus giving the body more time to eliminate it, still as unconverted DEG. DEG tastes sweet, and about a cupful is probably a lethal dose, but reports indicated that some survivors had consumed ten ounces, while some died from as little as an ounce.

Nobody knew any of this in 1937, but reports of people dying after taking the sulfa drug began almost immediately. Before long, animal studies confirmed that DEG was the culprit, and it was banned, but some 105 people had already died. The U.S. Congress passed the Federal Food, Drug, and Cosmetic Act in 1938. This required new drugs to be tested for toxicity before they went on the market, but while this might prevent the deliberate and ignorant use of DEG, it was not enough to stop further instances of DEG poisoning through carelessness or worse.

In 1986, 14 people died in Mumbai (then Bombay) when they received medications made up with glycerine contaminated with diethylene glycol. In 1992, there was a similar case in Nigeria, and, as recently as 1995, a terrible case in Haiti saw 60 to 80 children killed when glycerine mixed with DEG was used in an acetaminophen syrup. The adulterated glycerine originated in China and was shipped to Haiti from the Netherlands via a German company. It was labeled GLYCERINE 98 PCT USP, a claim that implies the glycerine is suitable for medicinal use, even though the shippers apparently were in a position to know it was only 53.9 percent glycerine.

As the events of 1937 showed, one of the problems in solving these cases is that the lethal dose varies widely from person to person, and the dosages taken may vary because many patients stop taking medicine as soon as there is an improvement in their condition. On top of that, there is always the possibility of genetic variation in the risk involved—as we have seen, Native Americans have less efficient alcohol dehydrogenase, and so may well have a better chance of eliminating DEG before it is converted to oxalic acid.

Another problem specific to DEG is that the victims of kidney failure do not present all at one time. Any number of diseases and contaminants can cause kidney failure, and the poison cases will be seen alongside disease-driven kidney failure, so it takes a subtle mind to identify the cases that are "wrong." Jagvir Singh and his colleagues outlined their investigation of a 1998 mass poisoning in India, when at least 33 children died. Researchers visited the houses of many of the dead children—all of whom had died of kidney failure—looking for any common factor. Naturally enough, they concentrated on obvious causes and effects. Blood and stool samples were taken from sick children, and water samples from tanks and tube wells.

The outbreak was centered in the Gurgaon district of Haryana state. Of the 25 children for whom treatment histories were available, 15 had been to the same health clinic, and while there were no records kept at the clinic, the qualified pediatrician said he had not changed any procedures recently. He had treated them with antibiotics, acetaminophen syrup, cough syrup, or rehydration salts. He treated about 100 children each day, and commented that it was odd so few had become ill.

The art of epidemiology is the art of finding the smoking gun, but here it seemed that maybe the firing pin had missed—there was something not right. The breakthrough came, say the researchers, when a paper in a medical journal described the outbreak of DEG poisoning in the Haiti case and also referred to the 1986 case in Mumbai. Investigators then realized that while most of the children would have been given the more commonly prescribed acetaminophen syrup, only a few would have been given the cough syrup.

Chromatography revealed that a locally manufactured cough syrup was 17.5 percent diethylene glycol. It also confirmed that none of the other prescribed medications contained any diethylene glycol. Some of the children who had been prescribed cough syrup had also been given injections of antibiotics. These can harm kidneys previously damaged or under attack, so in this case it required a particular, terrible combination to kill a patient.

Unlike in Haiti, where activists have been enthusiastic in blaming large and heartless pharmaceutical companies for the contamination, or in Spain, no blame seems to have been allocated for either the Indian or Nigerian outbreaks.

Other poisons that can cause disease are under investigation. In research conducted at Cornell University in 1998, pregnant rats were given drinking water contaminated with lead. The lead level was comparable with that in drinking water from some areas of the United States, and the rats produced offspring with damaged immune systems. If the same effects occurred in humans, fetal exposure to lead could account for the upsurge in asthma and other allergies, as well as cancers.

Those who have asthma and rely on "traditional" or complementary medicines face an extra hurdle. In 1975, 74 patients in Singapore were found to have arsenic poisoning. The source was a variety of anti-asthmatic herbal preparations, some containing up to 107,000 ppm of arsenic, or almost 11 percent! More recently, potentially toxic levels of arsenic have been found in traditional Chinese herbal balls sold in the United States.

Kill or Cure?

Historical analysis of famous deaths is a popular parlor game for medical practitioners, but it is hampered by the fact that our ancestors' lack of knowledge about microbes and infection led them to attribute any rapid death to poison. Extra confusion arises from the regular medical use and abuse of poisons in "cures," and the option for the cleverer doctors to keep quiet about some of their worst excesses. Not that they always did hide the facts—in some cases the doctor was sufficiently convinced that his treatment was fine; it was just that his patient had been uncooperative and died. In those cases the full report is available for us to examine in hindsight.

That alone is why we know that Philip II of Spain died in 1598 after enduring two months of hideous purging, and that Louis XIV had a close shave with death in 1658. In what follows, keep in mind that purging and bleeding were standard treatments until well into the nineteenth century, and taught in all the best medical schools. By themselves, they were terrible; in combination, they could be deadly. In some ways, it would seem, the poor were better off than the rich because they could not afford to be treated by doctors. Louis XIV was rich enough to afford the best and lucky enough to survive their ministrations.

At the age of 20, Louis was diagnosed with typhoid while on campaign in Flanders. First, his doctors tried bleeding him, but he developed a fever and convulsions and was in considerable pain. His doctors then debated whether or not to dose him with antimony, in the form of antimony wine. This was prepared by leaving wine to stand in a vessel made of antimony. The acidic wine slowly dissolved the metal, but the final concentration depended on the type of wine and the purity of the antimony. This was a remarkably risky preparation, and if luck had not been running Louis's way, history could easily have been rather different. As it was, he survived, and his kingdom flourished.

According to Dr. Johnson, our first lexicographer, antimony gained its name after a German monk named Basil Valentine gave an antimony supplement to pigs in their feed, and they thrived on it. Inspired by this, he fed the same mix to his brother monks, says the worthy doctor. The ungrateful recipients promptly sickened and died, causing the mix to be called *antimonachus* (or *anti-moine*—monks'-bane), meaning "bad for monks." A nice tale, even if untrue, but a fair indication that antimony was already recognized as unpleasant stuff in the mid-eighteenth century. Mostly, it was used in tartar emetic, a poison used to make the patient vomit and cast up whatever else might be troubling him or her.

Kings and queens were not the only victims of officious and official poisoners. In the United States, Benjamin Rush laid about him with bloodletting and calomel (mercury chloride), greeting the resultant mercurial fever as a sign the patient was recovering, and reassuring his students that it was very hard to bleed a patient to death. George Washington was forced to submit to the tender mercies of Rush disciples in December 1799, when he came down with a fever and sore throat.

First, local bleeders took three pints of blood, and the hapless patient was given two doses of calomel and a purging enema. Then another pint and a half of blood was taken, and the former president was dosed with 10 grains of calomel—enough to stop a fit man in his tracks. This was followed by several doses of tartar emetic, and a blistering compound was applied to the throat (blisters being supposed to draw off the harmful elements). Then, for good measure, he was prescribed a vinegar poultice, and the soles of his feet were blistered. When Washington managed to make himself heard, he, not surprisingly, asked to be left to die in peace. He died less than 24 hours after being taken ill, almost certainly a victim of his doctors' enthusiasm for poisons.

Abraham Lincoln, on the other hand, may have poisoned himself over a number of years. In 1858, Lincoln became so excitable during a debate that he grabbed a former fellow congressman, O. B. Ficklin, by the coat collar and lifted him from his seat, roaring, "Fellow-citizens, here is Ficklin, who was at that time in Congress with me, and he knows it is a lie." Legend has it he shook Ficklin until the man's teeth chattered. Fearing Lincoln would shake Ficklin's head off, Ward Hill Lamon grasped his hand and broke his grip. After the debate, Ficklin, who remained a lifelong friend, said, "Lincoln, you nearly shook all the Democracy out of me today."

Modern opinion is that Lincoln's volatility was caused by mercury poisoning. Lincoln's famous composure during his presidency could be explained by the fact that the behavioral effects of mercury poisoning can be reversed. Apparently Lincoln had taken a mercury pill called "blue mass" for some years to treat his persistent "melancholia," but in 1861, a few months after the inauguration, he stopped using it, saying it made him "cross."

These blue mass pills were compounded from licorice root, rose water, honey and sugar, rose petals, and 375 micrograms of mercury. Today, the safe maximum level for an average adult is 21 micrograms, and Lincoln would have taken two of his tablets at a time—40 times the safe dose.

Mercury was widely used not only by physicians but also by quacks, simply because, as a poison, it killed many diseases (as it did quite a few patients). Daniel Defoe is our witness:

'Tis sufficient from these to apprise any one of the humour of those times, and how a set of thieves and pickpockets not only robbed and cheated the poor people of their money, but poisoned their bodies with odious and fatal preparations; some with mercury, and some with other things as bad, perfectly remote from the thing pretended to, and rather hurtful than serviceable to the body in case an infection followed.

Daniel Defoe, *Journal of the Plague Year*, 1722

Strychnine could be found in at least one quack medicine, Fellow's Compound Syrup of Hypophosphites, which was sold as a "tonic alkaloid." Each fluid dram was said to contain 1/64 grain of strychnine. Easton's Syrup, on the other hand, contained quinine and strychnine, in addition to iron: it was supposedly an all-purpose tonic—though it sounds as though it might have had a secondary function as an abortifacient.

HARMFUL HERBS

Nature abounds in poisons, more of which we shall meet later. The traditional medicines we extract from animals, plants, and minerals often have a long history. Hippocrates recognized that willow bark—aspirin to you and me—eased aches and pains. Many of the older drugs in the pharmacopoeia are truly ancient. More importantly, many of them are poisons, but they have to be to kill the germs that infect the patient. In an ideal situation they are more poisonous to the germ than the patient.

From a modern perspective, some of the more traditional remedies seem just plain odd. John Hall, physician and son-in-law to William Shakespeare, quotes one cure that was probably as harmless as it was free of any effect. Then again, a hot onion in the crotch might just be unusual enough to shock people into urinating:

To the region of the bladder and between the Yard and the Anus was applied hot the next: Take a good big Onion: and head of Garlick, fry them with butter and vinegar. These thus used, procured Urine within an hour, with some stones and gravel . . .

John Hall, *Select Observations on English Bodies*, 1657

Onions, even when applied to one's private parts, are not poisonous to humans. They are poisonous to dogs, however, causing hemolytic anemia if ingested in sufficient quantities.

A poison could be defined as any substance that destroys the health or life of a living organism, but 100 cups of coffee, 250 grams of salt, and 200 kilograms of potatoes will all kill an adult human. Vitamin D is essential in small doses, but large doses will kill. As Alfred Taylor said, "A poison in a small dose is a medicine, and a medicine in a large dose is a poison."

A poison is a substance that interferes in some way with some process of biochemistry: it is an atomic- or molecular-level monkey wrench in the works. The toxicity of a substance will be demonstrated in the way it attacks the liver or the kidney, or blocks the delivery of oxygen to where it is needed, or the way an enzyme is either formed or operates. In the strict sense, a poison is a substance that can cause illness or death when taken in small amounts. In the legal sense, it is usually taken as a chemical with an LD_{50} of less than 50 milligrams per kilogram, but that leaves us in need of a definition of this term.

The LD_{50} is what we use in most cases to measure toxic effects. Shorthand for "the lethal dose required to kill 50 percent of a sample," it is defined more formally as the dosage, measured in milligrams per kilogram of body weight (the same as parts per million or ppm), required to kill 50 percent of a sample population within 14 days. Any substance, even something that is not legally a poison, can have an LD_{50}—table salt, sodium chloride, has a human LD_{50} of about 3 grams for each kilogram of body weight, while the LD_{50} of solanine, a poison found in potatoes, would be consumed by an adult eating 200 kilograms of potatoes in one hit.

It might be worth noting here that when we talk about LD_{50} values in humans, nobody has taken human populations and fed them varying doses of poison to see how many die. Instead, the values are inferred from case histories, one of the reasons why human LD_{50} values are generally stated as approximations. On top of this, some chemicals irritate without poisoning, but a sufficient irritation can lead to fluid in the lungs and so kill indirectly. Corrosives can also kill when they are swallowed, and, if the LD_{50} fits, they may be regarded as poisons.

Gases are a little different, and here we speak instead of the concentration–time combination needed to kill 50 percent, the LCt_{50}, which is going to be given as something like 20 minutes at 100 milligrams per cubic meter, or 10 minutes at 200 mg/m^3. In each case, the product (and so the LCt_{50}) is 2,000 mg min/m^3. We can extrapolate and say that, if the concentration is 10 mg/m^3, we would expect half of a sample to survive for 200 minutes, but some would die much sooner, and at some point the dilution will be so great that the body's defenses may be able to destroy all the poison as it is breathed in. Such is the cold mathematical precision of toxicology.

Modern medicines are generally designed to poison a microbe that is harming or poisoning us with some waste product. The value of the poison/medicine lies in it being more poisonous to the microbe than it is to us. Before there was a serious germ theory of disease, the assumption was that all disease was caused by some poison or another, and so the best treatment was to expel the poison by applying some form of emetic or laxative. (This was justified by doctors who could show that many illnesses and many poisons caused violent expulsion at one or more ends of the alimentary canal. In any case, people felt so wretched when being purged that they felt much better afterward—so purging just had to be good for them.)

At least as far back as Dioscorides, emetics were seen as appropriate, and indeed they still can be for some cases of poisoning, where there is no risk of the expelled poison getting sidetracked into the lungs and doing even more harm. All the same, we need to be a little careful in reading old recipes. John Hall mentions using agaric. He was, I hope, talking about a tree fungus from larches in the Levant, a common purgative. The same name was also used for fly agaric, which was mixed with water and put out as a fly poison, and also eaten to cause a form of intoxication. The Viking berserkers are thought to have eaten fly agaric to prepare themselves for battles, but Hall's agaric had no such interesting properties.

We will come back to Hall's remedies later, but doctors were not the only ones with poisons in reach. Gardens typically had a number of herbs with either known or reputed properties. A number of these poisonous substances were also believed to be useful if somebody wanted to expel an unwanted fetus. In short, they were abortifacients.

Abortion is, and perhaps always will be, an issue that arouses strong emotions. The Hippocratic oath included a promise not to administer abortifacients, but it is fairly clear that prostitutes in ancient Greece relied on abortion, and it must have been fairly common, since temple inscriptions indicate that a woman was regarded as being impure for ten days after being aborted.

According to the Stoics, a fetus was more plant-like than animal-like, and only became an animal at birth, so abortion posed no ethical problem for them.

In the age of the Roman republic, it appears to have been both acceptable and normal for women to rely on abortion, and both Galen and Dioscorides list many plants that could produce an abortion, taken either orally or as a vaginal suppository. Even the emperor Domitian, who died in AD 96, is said to have poisoned his niece, Julia, in an attempt to abort their incestuous child.

Then somewhere around the time of Severus and Caracalla, not too far from AD 211, abortion became a crime against the rights of parents, an offense punishable by temporary exile. The early Christians regarded abortion after the fetus was formed at 40 days as killing a living being, an attitude that was probably reflected in the later emphasis on emmenagogues, treatments taken not to abort a fetus but to restore a temporarily interrupted menstrual flow—the same thing by a different name, in many cases, since one way of restoring the flow was to abort the fetus.

The range of possible treatments was wide. Cantharides, which we will meet elsewhere as a fabled means of provoking unbridled lust, was used also to make good the damage that lust might wreak, but so too were colocynth, aloes, hemlock, saffron, pennyroyal oil, and juniper. There are at least 25, and perhaps as many as 50, species of juniper, but the abortifacient comes from just one of these, *Juniperus sabina*. This is the source of savin, or oil of savin, an effective killer of worms, and also, it appears, of fetuses. Traditionally, gin is flavored with juniper berries, and so it was thought to be an effective agent when a miscarriage was desired.

The name *gin* is derived from the French name for juniper, *genièvre*, and the juniper in gin is *Juniperus communis*. This yields oil of juniper, a powerful diuretic that is still in the British and U.S. pharmacopoeias. Unlike oil of savin, this appears to have no effect on the fetus. Yet another juniper, *Juniperus oxycedrus,* provides the cedar oil still found in most of the European pharmacopoeias but not in that of Britain.

This oil was a key ingredient in the discovery of bacteria in the nineteenth and early twentieth centuries. Much of what was discovered arose from the extra magnification seen when the object under scrutiny was attached to the cover slip of a slide by a drop of the strong-smelling oil. This arrangement became known as the oil-immersion lens, and it delivered decent resolution at a magnification of 1,000.

Juniperus communis

In that way, one oil of juniper has saved many, many lives.

Alfred Taylor was not convinced oil of savin actually caused a miscarriage, but speculated that it may have caused a shock to the system and so indirectly have procured an abortion. He noted that as long as the oil was used with this intention, it was likely to poison young women from time to time. Rough treatments like oil of savin aside, women traditionally grew a number of herbs that could, when applied the right way, relieve them of an unwanted burden. But how safe were these old remedies? Used correctly, they were still reasonably safe—and effective.

The aconite described by Dioscorides was leopard's bane, *Doronicum pardalianches*, a member of the daisy family, which he said was used in eye medicines. Listing the common names, Dioscorides identified monkshood, the modern aconite, as "the other aconite," and said it was the one used against wild animals. Dioscorides reported no medical uses for "the other aconite," which modern botanists call *Aconitum napellus*, and which is a member of the buttercup family. Later on, though, Galen would assert that both had the same medical properties. As we will see later with hemlock, common names can bring their own special problems.

In 1987, John Riddle listed 257 drugs used in ancient Greece and found that 230 of them still appeared in at least one pharmacopoeia, somewhere in the modern world. Of course, new discoveries have been made since classical times, like camphor, discovered in the sixth century on the Malay peninsula (or thereabouts). Camphor reached Europe by the ninth century, courtesy of Arab traders, and has been there ever since. Riddle also notes that, by 1979, 20 of the 25 most commonly prescribed drugs had been discovered since 1950, and this process has been accelerating. We still have our poison drugs, less familiar ones now, but no less toxic for that.

But if the medicine chest or the herb garden did not yield enough poisons, the boudoir had a few unpleasant items to offer. Fowler's Solution, a 1 percent solution of potassium arsenite, was a market leader in its time and so worth considering in a little detail. During the latter part of the eighteenth century, in the reign of King George III, this patent medicine was advertised as an "infallible remedy for agues and intermitting fevers" and able to work its miracles even where "the bark" (quinine) had failed. It was said to be derived from "cobalt," probably an arsenical mineral. The drops had the advantage that, unlike the incredibly bitter quinine, they were tasteless, and the remedy was sometimes used in hospitals, where it seems Dr. Fowler first encountered it.

He must have been impressed with what he saw, because in 1783 he asked an apothecary named Hughes to duplicate the preparation. The apothecary dutifully made up an alkaline solution of white arsenic that actually possessed therapeutic properties. In time, this became known, a little unfairly, as Fowler's Solution. It was first listed officially in the London pharmacopoeia of 1809, and it then became popular with women, who drank the solution for their complexions. Some used it as a cosmetic wash, and others even rubbed it into their hair and scalp to destroy vermin.

We will look at this in more detail in the next chapter, but while prostitutes commonly used arsenic for their complexions, women regarded as respectable did so as well. Elizabeth Siddal was first the model, and later the wife, of Dante Gabriel Rossetti. She became a great arsenic user, because she wanted to retain her youthful looks, bright eyes, and clear complexion. Unlike some of her poorer and less-educated sisters, she apparently knew that once she started she would not be able to stop. This may have something to do with her ultimate suicide from an overdose of medicinal laudanum in 1862, but even then she was denied eternal rest.

In 1869, Rossetti obtained permission to have his wife exhumed from her grave in Highgate Cemetery, as she had been buried with the only copy of some of his poems, verse that he now regretted having placed in such security. The Home Secretary gave the necessary permission, and the evil-smelling book was retrieved, disinfected, and later dried and copied. The poems were published in 1870, giving literary Victorians a subtle twinge of vicarious necrophilia.

When we look at some of the fashionable beauty treatments of today, we might be excused for wondering just how far, if at all, we have traveled from the nineteenth century. Let's take

Botox, botulinum toxin, as an example. On the one hand, the media warn us that this is a fearful substance that terrorists might well use to murder us in our beds, but women's magazines and wrinkle-free models and celebrities sing its praises—to such an extent that fashion-conscious teenagers as young as fourteen and fifteen are reportedly receiving Botox injections in cities like New York and Miami.

In simple terms, the toxin makes muscles relax. It eliminates frown lines by immobilizing the muscles that yield them. Botulinum toxin is a by-product of *Clostridium botulinum*, an anaerobic bacterium. In real life (or death, perhaps), the effect of the toxin is to kill nearby flesh, providing more oxygen-free living space for the bacteria.

The toxin blocks the release of acetylcholine from the endings of motor nerves. As the nerves cannot deliver signals, the muscles they would usually operate are effectively paralyzed. This paralysis lasts for some months, even up to a year, before the patients have to come back for another dose, or have a suddenly wrinkly face. It is a license for its purveyors to print money.

Botox was first used for strictly medicinal purposes, to treat lazy eye and uncontrolled blinking, but researchers soon noticed that wrinkles were also affected. A few side effects have been seen in some cases—minor allergic reactions, bruising, or a temporary drooping of the eye, but nobody has died of the tiny dose used.

Botox also has some rather more important cosmetic effects for those unfortunates who suffer from a form of excessive sweating, called hyperhidrosis. Sufferers become saturated, even when they use pads and antiperspirants, causing them intense emotional upset and social stigma. The quick fix is a Botox injection into the underarm skin, the palms of the hands, or the soles of the feet, where it paralyzes the sweat glands for six months to a year.

Botoxing is not really the thing to have done at parties by people who think *Ab Fab* is a documentary—after all, how many people would change their hair color at a "hair color party"? Nonetheless, there would seem to be at least a few valid reasons why some people may need small amounts of Botox from time to time, but it's not the only poison in the makeup bag.

6

COSMETIC AND
DOMESTIC POISONS

Mr. Pritchard, failed bookmaker, who maddened by besoming, swab-
bing and scrubbing, the voice of the vacuum-cleaner and the fume of
polish, ironically swallowed disinfectant.

Dylan Thomas, *Under Milk Wood*, 1953

The popularity of arsenic for cosmetic purposes seems to have
come either from the fabled "fair Circassians" or from the
Albanians, but it was by no means a craze limited to women's
cosmetics. Florie Maybrick soaked flypapers to remove the
arsenic they contained. She used the resulting solution as a facial
toner, and, as we saw in chapter 5, many patent solutions and
creams—for tightening the chin, removing freckles, or bright-
ening the complexion—contained copious quantities of arsenic.

Arsenic even had a role to play in death, whether or not it
was the cause. The embalmer's arsenic bleached skin to an
acceptable white and left the cadaver more supple, so it could
be posed more naturally. Embalming became popular in the
Civil War, when families wanted their war heroes brought home
from the battlefield for burial. The amount of arsenic used on

a corpse reportedly varied from "four ounces to 12 pounds," but even the lowest dose would have made it easier for poison murderers to escape discovery, so, over time, arsenic was replaced by formaldehyde.

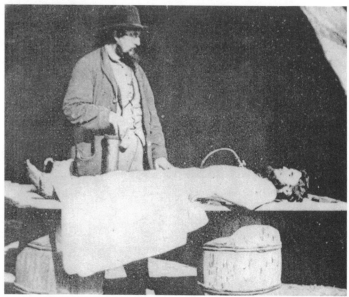

Dr. Richard Burr performing the embalming process on a body recovered from the battlefield.

Women and war dead were not the only users of arsenic: in chapter 2 we met two men whose partners were accused of their murders but who shared the curious habit of arsenic eating (arsenophagy). This is not quite as bizarre as it sounds, since we consume between 12 and 50 milligrams of arsenic each day in our normal diet. We can handle this because the amount of arsenic excreted in the urine each day is generally up to 50 milligrams. The Japanese, whose diet is high in fish and shellfish, have been found to have an arsenic content in their daily urine as high as 148 milligrams.

The men of the mining district of Styria, southwest of Vienna and centered on today's Graz, were renowned for their enthusiastic arsenophagy. They ate arsenic to improve both their skin and their breathing and to give them longer wind—most desirable in a mountainous area where a great deal of climbing was required. News of the practice appeared first in a Viennese medical journal in 1851 and soon after was being discussed in England, both among medical men and in the popular press. Arsenic became a core ingredient in preparations for everything from venereal disease to tapeworms.

While there can be no doubt that some Styrian arsenic eaters did indeed swallow large amounts of arsenic, quite large enough to poison a normal human, most of their intake probably passed out again without having been absorbed, because the grains of the mineral were too coarse to allow rapid uptake. So a lot of the arsenic eating may have been for show.

Arsenic was not the only poison found in cosmetic preparations. Belladonna, an extract of the deadly nightshade, could be applied to the eyes, where it would dilate the pupils. Despite the desirability of this "doe-eyed" look, it has become apparent only recently that women's pupils dilate when they are sexually interested in somebody, so the gentlemen were being led on, even when the women were completely unaware of what they were apparently "responding" to, or their "response."

Doctors also believed lead had some definite curative and cosmetic properties. John Hall noted that he used cerussa, otherwise white lead or lead carbonate, as a treatment for "pustles," or pustules, as we would say today:

> [Alice] Austin, a Maid, had her Face full of red spots, with red Pustles, very ill favoured, although otherwise very comely, and of an excellent wit. . . . The Body thus purged, her Face was anointed with the following Liquor: . . . Litharge of Gold powdered, one ounce. Alum one dram.

Borax three drams. Cerussa half an ounce. Vinegar two ounces. Rose-
water and plantain water, each three ounces. Boyl them to the wasting
of the third part, after strain them, and add the juyce of lemons, half an
ounce. . . . I advised her morning and night (the Pustles opened, broken
and crushed) she should wash the Pustles daily with the said Water.

John Hall, *Select Observations on English Bodies*, 1657

This would have been extremely painful with broken flesh, but
the pain probably made any cure seem even more effective. Hall
also used "litharge of gold," a yellow crystalline lead oxide, as a
face powder. In France, lead compounds were known as *poudre de
la succession* ("succession powder"), owing to their effectiveness
in removing annoying barriers between an individual and an
inheritance, but even that did not sound any alarms at first.

In Tudor times, those desirous of appearing in the forefront
of fashion might whiten their faces with white lead. Kohl, for
millennia a common cosmetic, is an antimony compound, but
one of the most bizarre uses of a poison must have been when
thallium was found to cause hair loss.

Thallium acetate was first given therapeutically to terminal
tuberculosis cases, in order to suppress night sweats. It may or
may not have worked, but the side effects of the treatment were
quite obvious—the patients' hair fell out. Accordingly, in 1898
the chief dermatologist at the St. Louis Hospital in Paris introduced thallium as a pretreatment for ringworm of the scalp.
After World War I, it was extremely popular for this purpose,
and Hamilton and Hardy record its use:

Thallium is absorbed through the skin, and so was used as a depilatory
to get rid of parasitic diseases of the hair follicles in children. If the
dose is right, children will completely lose their hair on the 16th to
18th day.

Asher Finkel (ed.), *Hamilton and Hardy's Industrial Toxicology*, 1983

Thallium acetate was sold in the United States as a depilatory under the name Koremlu Cream, until the mounting flood of damages claims for ailments ranging from neuritis to myalgia, arthralgia, and permanent loss of scalp hair drove the makers into bankruptcy. Even if you decided to keep your hair and dye it, you could be in trouble, as Mrs. Anna White was at pains to point out to her lady readers:

> Dyeing the hair is a very dangerous business, as most of the hair-dyes have for their base sugar of lead, caustic alkalis, limes, litharge and arsenic, all of which burn the hair. We have known of cases of paralysis of the brain occasioned by the inordinate use of hair-dyes which their makers asserted were 'perfectly harmless.'
>
> Shampooing is a great detriment to the beauty of the hair. Soap fades the hair, often turning it a yellow. Brushing is the only safe method of removing the dust from the head, with the occasional use of the whites of eggs. Perfect rinsing and drying should follow all washing of the head.
>
> Mrs. Anna R. White, *Youth's Educator for Home and Society*, 1896

The lead salts in hair dyes attach to disulfide bridges in the protein of hair to form black lead sulfide, destroying the bridge in the process. Lead in your system interferes with heme and porphyrin synthesis, which can lead to lower hemoglobin levels. Other symptoms of lead poisoning can include colic and abdominal pain, tiredness, and constipation, and even very low levels seem to affect intellect and learning.

Other salts used in hair dyes include copper, iron, nickel, cobalt, and bismuth, but women were not the only casualties of poisonous cosmetics. One American whisker dye was found to be as dangerous as the dyes offered to women because it contained excessive amounts of silver, and men also used whisker removers containing lead acetate, which could cause permanent baldness.

Mrs. White was full of good ideas for improving one's appearance. Lily-white hands had to be matched with lily-white complexions. She offered harmless remedies, such as bran mittens for keeping the hands white in spite of the disfiguring effects of housework. These were large mittens, filled with wet bran or oatmeal, that were to be worn overnight. This replaced a toxic seventeenth-century method of applying sorrel juice (containing oxalic acids or oxalates) to the hands. Freckles annoyed many ladies, she said, recommending Unction de Maintenon, which contained, among other things, oil of bitter almonds:

Venice soap, 1 ounce

Lemon juice, 1/2 ounce

Oil of bitter almonds, 1/4 ounce

Deliquidated oil of tartar, 1/4 ounce

Oil of rhodium, 3 drops

Dissolve the soap in the lemon juice, add the two oils, and put the whole in the sun till it becomes of ointment-like consistency, and then add the rhodium. Bathe the freckled face at night with this lotion, and wash it in the morning with clear, cold water, or if convenient, with a wash of elder flower and rose water.

Mrs. Anna R. White, *Youth's Educator for Home and Society*, 1896

Mrs. White possibly wasn't up to speed on her history when she named her preparation. Madame de Maintenon was the last mistress (and secret wife) of Louis XIV, whom we met in chapter 5. Louis and Madame had the dubious honor of presiding over one of the most poison-riddled courts in history. Mrs. White's freckle soap contained poison, cyanide to be specific, in the quarter ounce of oil of bitter almonds.

In murder fiction, the odor of bitter almonds is always associated with sodium, potassium, and hydrogen cyanide, but only 40 to 60 percent of people can detect this, and the real number is perhaps less, as few are willing to undertake the experiment. While cyanide's effects were known to the Egyptians 5,000 years ago, the first description of poisoning by almond extract was in 1679, and of cherry laurel water in 1731, both of them forms of hydrogen cyanide. The Swedish chemist Carl Wilhelm Scheele (we will meet him later as the discoverer of Scheele's Green) first isolated hydrogen cyanide (HCN), or prussic acid, from the dye Prussian blue in 1782. Four years later, he died of his own discovery when he broke a vial of prussic acid. The way of the chemist was never easy.

Prussic acid can be extracted from a variety of fruit seeds, and it seems likely its presence evolved because it made life unpleasant for any animals crunching the seed, while it did no harm to those merely swallowing the seed whole and carrying it away. In this way, the cyanide worked like the capsaicin in peppers—selecting those animals that would spread the seed most effectively, and with the least harm.

Unfortunately for almonds and a few other types of seed, cooks found oil of bitter almonds to be a most attractive addition to various meals. According to Taylor, the essential oil of bitter almonds contains a variable amount of HCN, rising as high as 12 percent, while almond flavor or essence of peach kernels contains one dram of the essential oil to seven drams of rectified spirit. In either case, the stuff is fairly toxic. Fifty milligrams of HCN—a bit less than a grain—is enough to kill, and a bit of quick figuring reveals that a one-ounce bottle of the essential oil would contain enough to kill about 60 people. Strong stuff to keep in a kitchen if the cook has a temper, but an even bigger threat in the Victorian kitchen came from nitrobenzene, which has a similar smell and taste. It gives confectionery the smell, but not the pleasant taste, of bitter almonds, and packs a far bigger kick.

Cyanide acts on the blood. It has its effect, in simple terms, by starving the body of oxygen, which cuts off the formation of adenosine triphosphate (ATP), an essential part of getting energy out of food in a form that cells can use. Cyanide victims have cherry-red blood, because the hemoglobin is saturated with oxygen that can go no further, and in the midst of plenty, the victim dies. Just 50 mg of hydrogen cyanide or 375 mg of a cyanide salt can kill a human.

Technically, cyanide is a rapid-acting poison that inactivates cytochrome c oxidase when the cyanide ion reacts with ferric (trivalent) iron in cytochrome c oxidase. This enzyme is involved in the final stage of the electron transfer chain, which delivers energy. Cyanide inhibits the electron transfer chain and stops ATP from forming. Because of this, there is a buildup of lactate, leading to acidosis, especially in the brain.

Chemistry in the body is mainly a matter of checks and balances, with chemicals formed in one process being broken down in another, so as to reset the state of something. At some synapses—links between nerves—a chemical called acetylcholine is used to carry the message from one nerve to the next. The synapse is then reset as an enzyme called cholinesterase wipes the acetylcholine out again. Some nerve gases and organophosphate insecticides block cholinesterase, so the acetylcholine keeps on triggering the second nerve.

Strychnine attacks another sort of synapse transmission, one relying on glycine. Glycine works to inhibit signals, but strychnine locks the glycine out, with the result that the nerves keep firing, reflex arcs fire over and over again, and a form of spastic paralysis sets in, followed by convulsions and paralysis of the respiratory muscles, so the victim dies. The muscular contractions of strychnine produce characteristic contortions of the body, arched backward so only the heels and the top of the head touch the ground, and on the face, a *risus sardonicus*, or, if you recall your Tom Lehrer, a hideous grin.

But while Lehrer's victim was said to have died with the spoon in her hand, there are other ways of taking in poison than swallowing it. We have about two square yards of skin, and some 90 square yards of lung surface, so it is possible to inhale some poisons and to absorb others through the skin. Nitroglycerin, for example, can be absorbed through the mucous membranes of the mouth, which is why it is placed under the tongue when it is used as a treatment for angina pectoris.

Taylor notes that the symptoms of nitrobenzene poisoning are remarkably similar to those of cyanide. In at least one case he examines, essential oil of almonds was at first believed to be the cause. A woman who was making pastry tasted the bottle of what she had assumed was oil of almonds. Finding it acrid, she spat it out. She was unwell for a long time but survived her encounter with nitrobenzene. In another case, a 13-year-old boy tested a bottle to his lips and died some 12 hours later. The household's cook also applied the bottle to her lips and became ill, but later recovered. The boy may have swallowed some of the liquid, as his stomach contents had the characteristic smell of the liquid when an autopsy was carried out. The bottle had been wrongly labeled as oil of bitter almonds.

Deadly Drapes and Other Domestic Poisons

Serious chemical discovery got under way in about 1780 and, after a slow start, took off in the nineteenth century, as people worked out ways to harness the new knowledge. Much of it related to colorful (that is, generally reactive and potentially poisonous) new compounds. One famous, although putative, victim of this knowledge was Napoleon Bonaparte, eventually released from his miserable exile by a lingering and painful death.

A later check of the deposed emperor's hair revealed that there was indeed arsenic in Napoleon's remains, but this does not mean he was given a lethal dose. On the other hand, a large number of people would have been willing to see him dead. Alive, he was a threat to peace, in case he escaped, just as he had on Elba, but somehow, people could not bring themselves to kill him. Perhaps nobody wanted to carry on the tradition of executing kings and emperors, not even despised ex-emperors.

Ambroise Paré claimed after the event that the sudden and fatal illness in 1534 of Pope Clement VII, one of the Medicis, was caused by him inhaling arsenical fumes from a torch carried by one of his attendants in a procession. Others claim that both the pope and Leopold I of Austria were killed by the fumes from arsenic-laden candles.

These claims are impossible to prove, but there is a sound basis for them. Researchers have known for some years that lead wicks in candles can release significant amounts of lead into a room. These candles tend to be long-burning scented and ceremonial candles. In one set of tests, carried out in 1999, the lead levels emitted reached as high as 65 micrograms per cubic meter ($\mu g\ m^{-3}$) against a U.S. safety standard of 1.5 $\mu g\ m^{-3}$. In the same year, lead wick candles were banned in Australia, but they were not banned in the United States until 2003.

Diehard Napoleonists are still convinced that some person or persons unknown actually murdered the emperor. The issue was raised in Paris at a meeting of the International Napoleonic Society on the eve of the 179th anniversary of Napoleon's death, May 5, 2000. In the 1960s arsenic had been found in hair purported to belong to the emperor, and now it was claimed that the FBI had run further tests in 1995, finding levels of 20 to 50 parts per million (ppm) of arsenic, where 1 ppm is more normal.

Contemporary reports of Napoleon's terminal illness mention that he complained of light sensitization, loss of hair, sleep problems, and neurological disturbances. All these are consistent with arsenical poisoning, and postmortem reports suggest he was still fat when he died, which is inconsistent with the official diagnosis of gastric cancer. The problem with the claims is that there is no proof that the hair specimens actually came from Napoleon, and important (and hard to miss) signs of arsenical poisoning, such as leathery texture of palms and soles of feet, were not present in Napoleon's corpse.

There were no less than eight doctors at the postmortem, and they agreed that extensive stomach cancer was present at death. More importantly, one of them was one Francesco Antommarchi, who attended Napoleon through his last illness. Antommarchi was a Corsican, who had been sent to Elba by Napoleon's mother. It is unlikely he was part of a British plot to poison his patient, and he surely would have noticed and reported symptoms of arsenical poisoning. Let us assume, however, for the sake of argument, that the various samples of Napoleonic hair are genuine and do indeed contain excessive amounts of arsenic: there are quite a few ways for the arsenic to have gotten there.

Beginning about March 1821, Napoleon was given tartar emetic to induce vomiting. This may have produced the opposite result—an inability to vomit—if the mucous lining of the stomach was already corroded. In April 1821, he was given an almond drink that would have contained mandelonitrile, a substance that can decompose to produce prussic acid. Finally, on May 3, 1821, Napoleon was given a relatively large dose of 10 grains of calomel, ostensibly to relieve constipation. Reaction of calomel and mandelonitrile, or the prussic acid it formed, may have produced mercury cyanide. As Napoleon was unable to vomit, this toxic cyanide could have been retained in his system until it killed him on May 6.

On the other hand, there are two perfectly feasible ways for Napoleon to have ingested a certain amount of arsenic, possibly enough to kill him, in complete innocence and with no plot intended. In 1856, a Dr. Boner alleged that Napoleon had been in the habit of taking arsenic as a precaution against being poisoned, and this was confirmed by others. A second, and more likely, possibility is that it came from the wallpaper.

In 1778, Carl Scheele announced the discovery of a new copper arsenate dye, and over the next few years a number of slight variations on the theme were discovered. This was at a time when most dyes were vegetable based and given to fading or washing out. The new mineral pigments held their hue far better—they were brilliant and colorful, and they caught on in a big way. By 1814, the Wilhelm Sattler Dye and White Lead Company in Schweinfurt began making a mixed copper acetate–arsenate salt that gave a beautiful green in paper, textiles, and confectioneries.

In no time at all, Schweinfurter green was the new green all across the civilized wallpaper-conscious world. Its big plus was that it did not turn gray when it was exposed to sulfides— important at a time when coal was a common source of heat. A couple of minuses were that wallpapers were attached with starch paste or, rarely, animal glue, and damp-proofing at the start of the nineteenth century was far from ideal. As a result, many rooms ended up lined with soggy paper smothered in poisonous Scheele's green, Paris green, or Schweinfurter green and with a good supply of adhesive to provide fungus food.

In chapter 7 we will encounter biomethylation as a defense against toxic metals. Many fungi have the ability to drive this reaction, which makes metals more accessible, and in the soggy rooms of the civilized and tastefully wallpapered world small biochemical reactions were started, as the fungi struggled to survive the arsenical realm they found themselves in. (The dyes were also in artificial flowers, carpets, furs, dress fabrics, and black stockings, but the wallpaper was the main problem.)

People realized there was a problem with the new colors fairly quickly, because in northern Europe people soon noticed how bedbugs died in rooms papered with the new designs. At first even more of this useful paper was sold, but then people began to notice a garlic odor and that those sleeping in such rooms got sick; some died.

Somewhere around 1865, according to legend, somebody dumped some Paris green on a potato patch in America and noted that it killed insects while leaving the plants alone. This started a tradition of arsenical pesticides.

By 1838, the Prussian government had banned the use of poisonous substances in wallpapers. Regrettably, other parts of what we now call Germany—and the rest of Europe, for that matter— were not so strict. It was, in any case, too late for Napoleon.

We have known some parts of the story since 1897, when Bartolomeo Gosio (1863–1944) showed how a fungus then known as *Penicillium brevicaulum*, now known as *Scopulariopsis brevicaula*, attacked the starch paste and sizing, excreting an arsenic compound he could not identify. It became known as Gosio gas, but while his name is recalled by specialists, and while there is a via Bartolomeo Gosio in Rome, he is little remembered today, except by some who suspect crib death or SIDS may be caused by Gosio gas.

Gosio was investigating a number of deaths that appeared to be caused by a volatile, garlic-smelling arsenic compound emanating from damp, moldy rooms. The wallpapers in these rooms were colored with arsenic-containing pigments, and Gosio isolated a number of microorganisms associated with the gas. His assistant, Biginelli, trapped some of the gas as a complex with mercury(II) chloride. After chemical analysis, he suggested it might be diethylarsine, $(C_2H_5)_2AsH$.

When Clare Boothe Luce was U.S. ambassador to Italy, she was poisoned by arsenic, though it turned out to be an accident that should have been kept under wraps. Luce was aware that her behavior was irrational, and reported to President Eisenhower that she felt as though she was drunk or drugged on several social occasions. This report got out when Eisenhower's press secretary added the snippet to one of Ike's news briefings. At no time did Luce say she had been poisoned, only that her behavior was being oddly affected by something.

Richard Helms of the CIA later established the presence of arsenic in the ceiling of the ambassador's bedroom—as was common with many Italian ceilings—and it was claimed her bed was uncanopied (actually, they said she was sleeping on a sofa). Apparently there was a laundromat on the floor above, which caused the ceiling to vibrate, making one wonder: if the ambassador was living in primitive conditions like this, stuck on a sofa under the laundromat, where had they put the hired help?

The American press was delighted, because the lady and her husband (Henry Luce, owner of *Life* and *Time* magazines) had long been labeled Arsenic and Old Luce. The idea of Arsenic being poisoned with arsenic was too attractive to be allowed to rest, but on this occasion, even given the Italian setting, there was no state-sponsored poisoning going on. Arsenic is merely a common feature of room surfaces in old Italian buildings.

The Gosio gas was finally identified by Frederick Challenger as trimethylarsine in 1945, completing research he had begun in 1931. So when the arsenic was found in Napoleon's alleged hair, the basic information was in place. Around 1980, David Jones gave a radio talk on the issue and wondered what color Napoleon's wallpaper might have been.

There is a certain sort of delightful person who reads books or listens to radio talks and then writes to the author, providing an interesting and key snippet, fact, date, or name. If you are one of those, know that you are valued, for your snippets are the thread that ties the pieces together, launching your chosen recipient in an entirely new direction. Blessed are the snippet-passers of this world, though these days we often find them on the Internet.

In those days before the Internet took off, Jones received a letter from one Shirley Bradley, who had inherited a fragment of Napoleon's wallpaper in a scrapbook. The wallpaper showed a green and brown star, but it may once have been green and gold, the imperial colors, and have faded. The fragment was found to contain arsenic, so if Napoleon was poisoned with arsenic, it is quite likely it was by mischance. If this is what happened, it may be a poetic form of justice, as it is widely believed that Napoleon was not above trying to poison others, especially King Louis XVIII.

In 1804, Louis XVIII of France was living outside Warsaw, in what was then Prussian territory. One day, one of the king-without-a-throne's servants, one Coulon, was offered 400 louis d'or to add some hollow, poison-filled carrots to the soup that was to be served to the exiled royal and his family. Coulon was loyal: he accepted the carrots but denounced the emissaries. The Prussian police, conveniently, were careless enough to allow them to escape, arousing suspicions then and later that Napoleon himself may have been behind the plot, as he had a fair amount of clout at the time in the locality.

The carrots were later found to be filled with a paste of white, yellow, and red arsenic. Napoleon's involvement was never proved, but during the Franco-Prussian War his nephew, Napoleon III of France, had no qualms about proposing that French bayonets be tipped with cyanide. Perhaps he was just joking, given that cyanide is also known as prussic acid, but in 1870 such ideas were treated with some horror. As we will see in chapter 9, this horror at the prospect of using poison in battle would not last.

It is interesting to consider whether we have more or less poison in our households today than we did a few generations back. In the nineteenth century, homes needed lye (caustic soda) to make soap, arsenic flypapers, arsenic powder for treating termite infestations, iodine, and copperas (iron sulfate) and bluestone (copper sulfate) for cleaning and scouring. Matches contained phosphorus, which was also found in cockroach poison. In the mid-nineteenth century, arsenical rat poison was replaced by phosphorus-based rat poisons, but these were equally effective in doing away with both human and rodent. There was lead in the paint and pipes, and arsenic in the wallpaper, while at least some of the dyed clothing was quite likely to be poisonous.

Today, we would expect to see antifreeze in cold climates, assorted caustic drain and toilet cleaners, oven cleaners, dishwasher detergent, bleach, insecticides, rat and mouse poisons, nail polish removers, paint thinners, disinfectants, mothballs, alcoholic drinks, tobacco products, and, in a few places, flaking toxic paint. For the most part, today's household poisons are clearly labeled and generally have distinctive additives—color, taste, and smell—that make them unsuitable for a stealthy attack.

Disposing of lead-based paints continues to be a problem where old homes are being renovated, but at least people are now aware of the problems. Equally, the disposal of poisons from the workplace is gradually coming under better control. Our past record, however, is not good, so far as toxic working and living conditions are concerned.

7

POISONED WORKPLACES?

Oh what can ail thee, knight-at-arms,
Alone and palely loitering?
The sedge has withered from the lake,
And no birds sing.

John Keats, *La Belle Dame sans Merci*, manuscript version, 1819

After gaining her verdict of "Not Proven," Madeleine Smith moved to London and married Pre-Raphaelite artist and designer George Wardle. She was welcomed into those circles, and when George Bernard Shaw met Lena Wardle later, he found her to be a pleasant woman, not at all sinister, he said. She and Wardle had two children but separated after 28 years of marriage. She went to America a few years later and eventually married again and lived in New York, where she died in 1928.

This is not quite as surprising as it might first appear. Poison—real as well as figurative—seems to feature quite a lot in the arts, a notion that would surprise very few who have ever been involved even peripherally with art critics or arts administrators. Madeleine probably met Elizabeth Siddal among the Pre-Raphaelites, at least fleetingly, but she would have known many others in her husband's circle of acquaintances who worked with poisons every day. Many of those she knew were, in all probability, at least slightly poisoned by their materials and their studios.

Painters generally use mineral-based pigments for the same reason that Schweinfurter green was so popular. Vegetable dyes fade within only a few years, while minerals are usually much more stable. They might weather and break down in the field, but those bound up in oil on canvas are well able to survive in the shelter of the average home or art gallery. There is just one small catch: as a general rule, colored inorganic compounds are toxic, and artists must risk being poisoned to create great and lasting art.

One of these poisonous colors was *minium*, but its contents changed over time. The ancient Romans obtained their *minium* from Spain. The name came from the Basques, who used this word for red mercury sulfide. Others called it cinnabar, a Persian name that came into Latin via the Greeks. Minium came into the English language later, but by then the name was attached to red lead, Pb_3O_4, a useful color when red letters were required to mark something special—a rubric, from the Latin for "red." The verb for doing this sort of illustration was *miniate*, and the result was a *miniature*. Of course, the same practice also gave us "red letter days."

Chromium compounds, including lead chromate, yielded oranges, yellows, and some greens; cadmium compounds covered the reds and additional yellows; white lead, otherwise known as lead oxide, gave a nice matte white surface; while vermilion is just cinnabar by another name. Naples yellow contained lead and antimony, and so the list progresses, a spectrum of heavy metals. (Some mineral pigments are less stable than others, however. The white lead base has proved a problem in some old masters, because it slowly combines with atmospheric pollutants that contain sulfur, changing to black lead sulfide.)

Unscrupulous apothecaries might mix brick dust with vermilion, and so turn a handy profit, or they might sell azurite as the far more precious ultramarine, so the early artists needed to be passable alchemists, buying and grinding minerals to fine powders, making sure they were the right minerals, and concocting systems to carry them and hold them on a canvas. Even today, some chemistry is needed: for example, Prussian blue breaks down in alkali, which means it cannot be used in an acrylic paint. With all the dust that was blowing about in their studios, it is hardly surprising that painter's colic, *colica pictorum* to the Latin-favoring doctors, was known as far back as the eighteenth century. In 1767, Benjamin Franklin mentions having seen a list of tradespeople suffering from colic in Paris:

> I had the Curiosity to examine that List, and found that all the Patients were of Trades that some way or other use or work in Lead; such as Plumbers, Glasiers, Painters, etc. excepting only two kinds, Stonecutters and Soldiers. These I could not reconcile to my Notion that Lead was the Cause of that Disorder. But on my mentioning this Difficulty to a Physician of that Hospital, he inform'd me that the Stonecutters are continually using melted Lead to fix the Ends of Iron Balustrades in Stone; and that the Soldiers had been employ'd by Painters as Labourers in Grinding of Colours.
>
> Benjamin Franklin, letter to Benjamin Vaughan, 1786

As recently as 1962, the brilliant (if slightly mixed-up) Brazilian artist Candido Portinari died of lead poisoning from the lead-based yellows and whites he used. In 1948, he had used an arsenic-based paint and ended up in hospital. This is not why

I call him mixed-up, though—Portinari was an atheist who painted saints, a communist who painted the official portrait of a dictator, and an excellent artist to boot.

These days, artists are generally more aware of the poisons that surround them. There is evidence that such an awareness has been around for a while, but it was not enough to safeguard Raphaelle Peale, an important American still-life painter of the early nineteenth century. Peale's father owned a natural history museum where Raphaelle was the taxidermist. His father believed Raphaelle's physical and emotional problems were caused by gout and drink, but a combination of arsenic and mercury poisoning seems more likely. Yet Raphaelle was well aware of the dangers of the arsenic that was used to protect the stuffed birds from pests, and posted signs in the museum reading, "Do not touch the birds, they are covered with arsnic {sic} Poison." He lived from 1774 to 1825, which was not excessively bad in those days, but not outstanding, either.

The dangers of working with lead were well-known as far back as Roman times. The Roman writer Vitruvius mentions them, and it is significant that the patron god of metalsmiths was Vulcan. Like their patron, Roman metalsmiths were often lame, pallid, and wizened, but it was lead that made them so, not a kick from one of Vulcan's parents (there are conflicting versions on whether it was Jupiter or Juno who lamed Vulcan). This would also explain why lead was the metal associated with Saturn in Roman mythology. The association was with both the god and the planet, and so we sometimes come across references to "saturnine poisoning," which is just lead poisoning with social-climbing tendencies.

While the Romans agreed that lead was dangerous, they seemed to think small doses did no harm. The slaves who mined the lead might die—not those who ingested it through their food and wine. This may have been selective blindness, a bit like the attitude of Victorian industrialists to the plight of their poisoned workers, but neither case can be excused.

Hippocrates described a case of lead poisoning in 300 BC, and Dioscorides left us a detailed account of lead poisoning and the paralysis it caused. In it, he mentions lead fumes, and refers to *molybdania*, probably litharge (lead monoxide) and *minium*, red lead oxide. By the nineteenth century, lead was taking on an altogether more sinister role, at least in the industrial nations. Artists and philanthropists such as Robert Sherard and H. G. Wells spoke out about the evils of lead, even as everybody else tried to ignore them.

Sherard was a journalist and social campaigner who wrote for *Pearson's Magazine* and later published his work as *The White Slaves of England*, where he looked at some of the degrading trades that English workers were subjected to. We do not have the space to describe the plights of the nailmakers, slipper-makers, tailors, woolcombers, and chainmakers, but let us spend some time with the alkali workers and the white-lead workers, for here we will see a clear and dreadful picture of what it was like to work with poison.

Sherard got very seriously up the noses of the industrialists who believed it was their social and moral duty to make obscene profits. Like today's economic rationalists, they believed all social ills could be remedied if a few people became vastly wealthy. As they spent their money, everybody would benefit from their largesse trickling down the economic chain. The only flaw in their master plan was that for them to become very rich in the first place, a lot of people needed to stay very poor, very diseased, or very dead. People like Robert Sherard were determined to expose them for what they were. He was unwelcome— in fact, in *The White Slaves*, Sherard shares some of the viler attacks made on him for "scurrilous dickturpinism," libel, and worse. The factory owners were unlikely to give such a trouble-maker easy access to their workers, and a foreman told Sherard there had been too much written about them.

> The lead works are carefully guarded. High walls surround them like prisons, whilst each entrance is watched by yard policemen. The things that are done here are not for the public eye. A stranger could more easily obtain permission to visit the Tsar's palace of Gatschina than a *laissez-passer* for most of the factories in Newcastle where white-lead is made.
>
> Robert Sherard, *The White Slaves of England*, 1897

Nonetheless, Sherard managed. On one occasion, he interviewed the doctor who looked after the factory workers. The doctor had no doubt about the effects of lead:

> For checking a too rapid growth of the population, indeed, nothing better could be devised than the employment of women in the white-lead factories, for the lead woman—according to Doctor Oliver's long experience—almost invariably miscarries, while if the children are born, very few of them live.
>
> Robert Sherard, *The White Slaves of England*, 1897

The dangers of lead for pregnant women and their babies are borne out by evidence from Italy, gathered in 1930. In Milan, where miscarriage rates were around 4 to 4.5 percent for the whole population, the wives of printers showed a miscarriage rate of 14 percent, and women printers an appalling 24 percent. In 1930, the average death rate for the first year of life was 150 per 1,000; for the babies of women associated with printing, it was 320 per 1,000. A few years earlier, in Japan, a researcher had found that male lead workers in storage battery plants had 24.7 percent sterile marriages (against 14.8 percent for a non-lead group), with 8.2 percent of pregnancies ending early or in stillbirth, against 0.2 percent in a control group.

Sherard describes how workers would be examined for signs of plumbism, lead poisoning, and ordered away from work if the telltale signs presented.

> . . . there are certain signs which unmistakably betray lead-poisoning, and the most important and sure of these is the appearance of a blue line on the gums. This blue line, by the way, may be noticed in about 75 per cent of lead-workers. It is due . . . to the action of sulphuretted hydrogen upon lead circulating in the blood, and has been noticed well-marked in girls who have only worked one week at the trade.
>
> Robert Sherard, *The White Slaves of England*, 1897

Those laid off would do the reverse of malingering to get back to work, even applying to another works under a false name, because their own name would be on the blacklists circulated by the factory owners. Given the choice between a quick death by starvation and a slow death by poisoning, most opted for the poison.

One woman, Sherard said, told him that dangerous as the lead work was, it was either that or go on the streets, "and we prefer anything to dishonour." But powerful as his arguments were, he was mostly preaching to the converted, to those already aware of the evils their society imposed on its weaker victims. It needed a popular writer of fiction, such as H. G. Wells, to reach a wider audience. In his powerful *The New Machiavelli*, Wells described lead palsy in such terms that the housewives of Britain began to demand "fritted ware," rather than the traditional pots and crocks. To make a fritted glaze, the lead is added before fusing and forms largely insoluble lead silicate and lead borosilicate. The risks to both the makers of the crockery and its users were considerably lessened.

Wells's finely honed prose exposed the factory owners and their attitudes to the public's gaze. This was not the dickturpinism that Sherard was accused of—this was dickensism, pure and simple. Take Richard Remington, for example, a future politician, who is learning about the real world from his uncle, who owns a pottery where plumbism was rife.

> 'None of your gas,' he said, 'all this. It's real, every bit of it. Hard cash and hard glaze.'
>
> 'Yes,' I said, with memories of a carelessly read pamphlet in my mind, and without any satirical intention, 'I suppose you MUST use lead in your glazes?'
>
> Whereupon I found I had tapped the ruling grievance of my uncle's life. He hated leadless glazes more than he hated anything, except the benevolent people who had organised the agitation for their use. 'Leadless glazes ain't only fit for buns,' he said. 'Let me tell you, my boy—'
>
> He began in a voice of bland persuasiveness that presently warmed to anger, to explain the whole matter. I hadn't the rights of the matter at all. Firstly, there was practically no such thing as lead poisoning. Secondly, not everyone was liable to lead poisoning, and it would be quite easy to pick out the susceptible types—as soon as they had it— and put them to other work. Thirdly, the evil effects of lead poisoning were much exaggerated. Fourthly, and this was in a particularly confidential undertone, many of the people liked to get lead poisoning, especially the women, because it caused abortion. I might not believe it, but he knew it for a fact. Fifthly, the work-people simply would not learn the gravity of the danger, and would eat with unwashed hands, and incur all sorts of risks, so that as my uncle put it: 'the fools deserve what they get.' Sixthly, he and several associated firms had organised a

simple and generous insurance scheme against lead-poisoning risks. Seventhly, he never wearied in rational (as distinguished from excessive, futile and expensive) precautions against the disease. Eighthly, in the ill-equipped shops of his minor competitors lead poisoning was a frequent and virulent evil, and people had generalised from these exceptional cases. The small shops, he hazarded, looking out of the cracked and dirty window at distant chimneys, might be advantageously closed.

H. G. Wells, The New Machiavelli, 1911

Wells slips in a reference to a "sickly-looking girl with a sallow face, who dragged her limbs and peered at us dimly with painful eyes. She stood back, as partly blinded people will do, to allow us to pass, although there was plenty of room for us," obviously assuming his readers would know that lead poisoning can lead to blindness. The damage was to the nerves, and a British study in 1910 showed 7.7 percent of women potters with lead poisoning went blind, while 10.2 percent had some significant loss of vision, and 14 percent a slighter loss.

White lead used to be made by the Old Dutch process, where strips of lead metal are buried in spent tanbark (or horse dung) and doused with acetic acid. Lead acetate forms, and as the tanbark ferments it produces carbon dioxide, which converts the acetate to the carbonate. It was not until the 1940s that British Australian Lead Manufacturers (now Dulux) developed a safer way of making white lead, by exposing finely divided lead in rotating barrels with acetic acid and horse dung and tanbark, but in the bad old days the workers needed to remove the strips of lead from their acidic mulch by hand and scrape off the white lead crystals.

The chemistry is fairly straightforward. The horse dung and tanbark fermented, producing heat that evaporated the acetic acid, which reacted with the lead to make lead acetate. The fermentation also produced carbon dioxide, which reacted with the lead acetate to make white lead.

Those who jumped from the frying pan of England's factories into the fire of World War I experienced their own curious form of poisoning: recruits from lead pottery areas suffered a typical lead colic as a result of an exhausting military drill. The lead stored in their bodies had been released by stress and made them ill. After the war, it would be noted that the stress of infection or alcoholism could trigger acute lead poisoning in exposed workers, and cases of acute lead colic commonly appeared on Mondays after weekend excesses, but this was wartime, and normal exertions were damaging soldiers who needed to be fit enough to go over the top, and take their dose of ballistic lead—but here they were, complaining of colic! If it was going to affect the war effort, Something Had To Be Done.

By 1916, the use of lead in potteries had finally become the subject of official concern. Four classes of glaze were established as standards. A "leadless" glaze contained less than 1 percent lead (calculated as metallic lead); the next class had less than 2 percent soluble lead monoxide; the next less than 5 percent soluble lead monoxide; and the last group gave no account of their lead content and solubility (and presumably included those with more than 5 percent lead).

In 1916, Sir William Tilden reported with some pride that the British government was purchasing only leadless glaze, or glaze in which all of the glaze lead was "mainly in the insoluble condition," that is, fritted ware. This specification applied to telegraph insulators, glazed bricks and tiles, and sanitary and domestic ware, and it had a marked effect in driving manufacturers to meet the required standards.

After the war, however, a new problem had to be overcome: swords had to become ploughshares once more, as battleships were laid off and cut up during 1921. Between their gray paint and the steel, there was a protective coat of red lead, but if the steel had been protected at sea, the workers in the breaking yards were not. This red lead episode was over before the red lead's cumulative effects were sufficient to energize official intervention, but lead would play its part in a brand-new era of motor transport. In 1921, the same year the battleships were disappearing, Thomas Midgeley, Charles Kettering, and Thomas Boyd found that adding tetraethyl lead (TEL) to motor fuel reduced engine knock. The Ethyl Corporation was soon formed as a GM subsidiary, and the automotive industry took off, especially in the United States.

More roads, wider roads, better roads opened up, only to be filled with more and more vehicles, all spewing out the lead that had been added to their fuel. Market gardens near highways grew lead-spattered lettuce, people living near roads breathed fine particles of lead—it was everywhere. By 1924, the first warning sign appeared when Midgeley himself fell ill from working with TEL, but it was not sufficient to ring alarm bells.

By 1925, health checks on garage workers who manually added the TEL to fuel showed that the exposed workers were adversely affected, but chauffeurs and other garage workers seemed to be unaffected. It was agreed that the gasoline should be supplied already blended, and that the containers should carry warning signs, indicating that their contents were unsafe for cleansing and dangerous if spilled on the skin.

In 1926, a Surgeon General's inquiry was rushed through in seven months. It reported that there were no good reasons to ban the use of TEL, provided there were proper regulations, and it would be almost 50 years before lead levels would be grudgingly reduced, microgram by microgram.

Robert Sherard identified other poisons too, such as chlorine. Spring, he wrote, never came to Widnes or St. Helens, the centers of the alkali trade, because the foul belching gases killed every tree and blade of grass for miles around. Farmers complained so often of their crops being ruined by fumes that the factory owners found it cheaper to buy up the affected land than to pay for the damage. Trees cannot live in this wasteland, Sherard said, "but men must and do."

Surprisingly, the evidence against chlorine was hard to find. Dr. O'Keefe, who treated the St. Helens workers for over 30 years, told Sherard the alkali trade was an unhealthy one, even though the statistics failed to show it. "The chemical yard only kills a man three parts out of four, leaving the workhouse to do the rest." The good doctor conceded that sulphuretted hydrogen was "a terribly poisonous gas, and but one of several which in these alkali works shorten life."

He tried to look on the bright side. "There is this, however, to be said in its favour, that if it poisons men, it poisons microbes also, and its effect is to minimise contagion by fever. We have but three patients at present in the fever hospital." Dr. O'Keefe did, however, note that few men got above 60 years of age. It is difficult to see a bright side in conditions such as these:

> Roger is their best joke, as Roger is their worst enemy. Roger is the chlorine gas, which, pumped on to slaked lime, transforms this into bleaching powder. Roger is a green gas, and is so poisonous that the men (packers) who pack the bleaching powder after the process into the barrels in which it is exported work with goggles on their eyes and twenty thicknesses of flannel over their mouths, these muzzles being tightly secured by stout cords. They can pack but a few minutes at a time. A 'feed' of this gas kills its man in an hour.
>
> For all that, Roger is the butt, not the bogey. True, that at the cry 'Roger is coming! Clear lads!' so frequently heard in the works, a wild

sauve qui peut of panic-stricken men may be seen scurrying before a green, perceptible, and palpable fog borne on the wind, but all the same, once the danger is past, Roger evokes smiles.

Robert Sherard, The White Slaves of England, 1897

To be able to joke about "Roger" makes it clear that these men had little else to laugh about, but they retained a clear vision and black humor about what was being done to them. One of the men explained to Sherard how easy it was for the factory owners to manipulate the statistics. "It's like this. You get gas. We run to the office for the brandy bottle and say, 'So-and-so's got gas.' Brandy is served out. You go home and die. Doctor says you died of faint, and the proof is that brandy was needed to revive you."

We will return to chlorine when we examine its use as a calculated weapon of war. It was the same gas, just with a different motivation. For now, however, let us close with one of Sherard's finest lines:

In Wiston Workhouse is a legless man, with whom an armless man keeps company. They were both alkali workers.

Robert Sherard, The White Slaves of England, 1897

One of the saddest and most unnecessary forms of profit-hungry poisoning was the condition commonly known as "phossy jaw." It was seen first in the manufacture of Congreve matches—matches named after the famous Congreve incendiary rockets and packed with white phosphorus. These fire-starters were also called Lucifers, though Lucifer was originally the name of a potassium chlorate device with no phosphorus.

There was a long gap between the first reported case of phossy jaw, that of Marie Jankovitz in Vienna in 1838, and the eventual phasing out of the white phosphorus matches in 1906. The first

London case was recorded in 1846, and Boston's first case was recorded at about the same time. Cases were reported regularly thereafter, but not one of them need ever have happened, if a small invention of 1824, a clever use of catalysis, had taken off:

> Amongst the ingenious novelties of the present day is a machine . . . for the purpose of producing instantaneous light; which appears to be more simple, and less liable to be put out of order, than the Volta lamp, and other machines of a similar kind. It has lately been discovered that a stream of hydrogen gas, passing over finely-granulated platinum, inflames it. The whole contrivance, therefore, consists in retaining a quantity of hydrogen gas over water; which is perpetually produced by a mixture of a small quantity of zinc and sulphuric acid, and which, being suffered to escape by a small stop-cock, passes over a little scoop, containing the platinum, which it instantly inflames. From this a candle or lamp may be ignited . . . it forms an elegant little ornament—of small expense, and easily kept in order; and, once charged, will last many months.

> *The Gentleman's Magazine*, September 1824, p. 259

Matches, however, were cheap, as were workers, and platinum was expensive—and to ignite the hydrogen, nothing else would really do. So matches needed to be made, but the phosphorus problem could have been dealt with as soon as it surfaced, if anybody had cared. By 1845, the safer red allotrope had been identified, and by 1855, a Swedish researcher had developed a match based on this safer form.

By 1898, phosphorus sesquisulfide, P_4S_3, was being used for matches in France, and finally, in 1906, the Berne Convention banned the use of white phosphorus. Only the United States failed to sign this convention, citing constitutional grounds, but in 1913 the U.S. Congress imposed a punitive tax, through the Esch Law, on white phosphorus matches and so eradicated them by making them as expensive as phosphorus sesquisulfide matches.

We should not be excessively harsh on the manufacturers. After all, phossy jaw killed relatively few. It started usually with a headache and was an extremely painful condition. There was a foul fetid discharge from the jaw and sufferers had such "garlic breath" that they were shunned by their fellows. It caused a raging thirst, led to grotesque disfigurement, and was a chronic condition. It might not have killed many, but most of its sufferers might have welcomed death.

Phossy jaw was called "the disease," "the flute," or "the compo," from the composition paste used to make match heads. Each year, composition paste in Britain absorbed 60 tons of phosphorus, and the workers absorbed an unknown amount of this mass of poison. More importantly, individual factories varied widely in their ventilation and in the mortality rates of their workers. We must assume that there was either a different morality in those days of stern Victorian scrupulous church-going, or that the factory owners were remarkably myopic.

In 1862, Britain's Privy Council commissioned a report on match factories that recommended better ventilation and hygiene and separate rooms for workers to eat in, but this was not enforceable by law, and its recommendations were largely ignored. There could be little doubt phosphorus was a poison: the lethal dose of white phosphorus for an adult is about the same as in a box of matches.

In Britain phosphorus was used to murder family, friends, enemies, and potential postmortem benefactors until 1963, the year in which Rodine (a paste of bran and molasses containing 2 percent phosphorus, and the most common phosphorus poison) was banned under the Animals (Cruel Poisons) Regulations. It was none too soon, whether for animals or humans.

Mercury poisoning is also hard to miss, and mercurialism was known to the Romans as a disease of slaves, because only slaves were used in the Spanish mercury mine of Almaden. The emperor Justinian considered a sentence to work there tantamount to a sentence of death, and so unpleasant was the death that Plutarch criticized a mine owner who used slaves in mercury mining who were not convicted criminals.

The subject was raised again in the 1500s, when Andrea Mattioli described mercurialism, and another Italian, Giovanni Scopoli, described the condition at Idria in 1761. By then, some had been trying to counter mercurialism for just under a century: in 1665, a short working day had been instituted for Friulian mercury miners, who had to work just six hours a day. This was the first known law to enforce industrial hygiene anywhere in the world. It took over 200 years for this initiative to be finally adopted at Idria, in 1897.

Most people know mercurialism as a disease of hatters, though the connection is a curious one, known long before Tenniel drew his famous Mad Hatter for Lewis Carroll. Hatters used the mercury in the felting process, to make the fur fibers stick together to form felt.

There is a pretty legend that Saint Clement, later the patron saint of hatters, was on a pilgrimage to Jerusalem. He is said to have lined his sandals with camel hair to ease his feet and, over time, the combination of heat, sweat, and pressure formed a sheet of felt, which was transferred to the head as hatting. This may be news to the makers of Kyrgyz *yurts* and Mongolian *gers*, both traditional felt tents.

I doubt that Saint Clement's method would have been used to make hats, given the combined odors of camel and human sweat that would have impregnated the felt, but we do know that even before 1685 hatters in Paris were using mercuric nitrate in a process they called carroting, because it gave an orange color to the fibers. The process was a trade secret within the hatters' guild, because the key to persuading the fibers to hook up to each other is to roughen them. This makes the fibers limp and twisted, so as they are cut and blown onto a cone-shaped form and then pressed down with a hot wet cloth, they stick to each other.

The mercuric nitrate solution is still known in France as *le secret*, and the process of felting with mercuric nitrate *le secrétage*. After the revocation of the Edict of Nantes, in 1685, thousands of Protestants, Huguenot hatters among them, were forced out of Paris, many of them taking refuge in England. The use of the solution persisted across Europe into the early twentieth century, and even later in Russia, but there is now a nonmercurial method of felting, and, with luck, mercury has been phased out everywhere.

In the days when mercury vacuum pumps were used to

exhaust the air from lightbulbs, spillages were common and mercurialism was rife in the lightbulb industry. Mercuric fulminate was used to set off explosions in guns and rockets, while calomel (HgCl) was used in tracer bullets. Both caused quite a few problems for munitions workers in World War II, although nothing near as nasty as those experienced by an unfortunate Allied airman who was struck by a phosphorus tracer bullet. The missile dissolved in his body and poisoned him.

Most metals used in industry have their own special quirks. Antimony is used in alloys, including the lead in storage batteries, paint, glass, pottery, varnish, and tartar emetic—a poison that provokes vomiting and so sometimes negates other poisons. Large amounts of arsenic are used each year, not only as a poison in herbicides but also as animal feed additives.

Poisonous substances sometimes only act as poisons when they are applied in the right way. If a child bites on a thermometer and swallows the mercury, no great harm will be done by the metal (the glass may be more of a worry), as it will pass straight through and out of the body. On the other hand, the same amount of mercury in the form of mercury salts would be absorbed far more effectively and be far more dangerous. If the same amount of mercury went down the throat each day, there would be a progressive buildup. Surprisingly, the contents of the thermometer are far more dangerous if they are spilled and go into a carpet or floor cracks, as this will establish a low but dangerous level of mercury vapor in the air. About the only thing that would make the situation worse would be to try and clean the spill up with a vacuum cleaner.

About this time, the thoughtful reader with dental fillings may be wondering about dental amalgam. For about a week after getting a silver amalgam, the patient has raised mercury levels in the urine, but it is always within safe levels and falls away. In simple terms, the amalgam is a potential threat to dentists and their assistants if it is used carelessly, but not to patients. This has led to three court cases seeking damages from the American Dental Association being thrown out in 2003 as having no merit.

In one of the cases, a New York Supreme Court judge dismissed two of the suits against the American Dental Association because the complaints showed no "cognizable cause of action." The ADA hailed this, not surprisingly, as "a victory for dentistry and science over superstition and hearsay."

One of the most deadly of poisons, in terms of the number of its victims, is an alkaloid that is legally sold all over the world without any permit being required. Charles Lamb knew it for what it was:

> Stinking'st of the stinking kind,
> Filth of the mouth and fog of the mind,
> Africa, that brags her foyson,
> Breeds no such prodigious poison,
> Henbane, nightshade, both together,
> Hemlock, aconite—nay, rather,
> Plant divine, of rarest virtue;
> Blisters on the tongue would hurt you.
>
> Charles Lamb, "A Farewell to Tobacco," 1811

There can be few lethal substances as easily purchased as nicotine, and few poisons that bring in as much revenue to governments, though sometimes this may be a poor bargain: in 1993, China's tax revenue from tobacco was estimated at 41 billion yuan, while the estimated costs of smoking illness and death totaled 65 billion yuan. This, of course, is an underestimate, because the true costs of today's smoking will only show up in two or three decades.

In 1995, a quarter of all deaths in the United States and 14 percent of the deaths in Europe were smoking-related, but, in theory, the effects should be far higher. A single cigarette contains sufficient nicotine to kill a person if it were extracted and injected, yet no smoker dies of acute nicotine poisoning, because of the way it is taken in and filtered by the lungs. Direct skin contact is a different matter.

In one nineteenth-century case, a smuggler covered his skin with tobacco leaves in order to defraud the revenue. The leaves were moistened by his perspiration and he developed all the symptoms of acute nicotine poisoning. Even today, itinerant farm workers in North Carolina develop a form of acute nicotine poisoning, green tobacco sickness, each year.

The modern condition typically occurs after pickers are exposed to wet tobacco leaves, either early in the morning or after rain. Being an alkaloid, the nicotine in the leaves is rapidly absorbed through the skin. As an industrial disease, the condition was first recorded in 1970, and is exacerbated during the harvest by a process called priming, when workers break off the ripe leaves and hold them under their arms as they move down the rows. As the day progresses, say researchers, the workers' shirts and skin grow stiff with sticky—and poisonous—tobacco juice.

The region known medically as the axilla, described quaintly in one research study as "the area under the shoulder joint" and known in more robust circles as the armpit, absorbs more chemicals than other skin areas. It might be worth publicizing this condition more widely, both to alert workers to the problem and also to alert smokers to just what it is they are putting in their mouths, and where it has been.

TOXIC LANDSCAPES

So long as there is enough of a poison, it can kill, even a poison that is a natural part of the atmosphere. Take carbon dioxide, for example. In August 1986, Lake Nyos in Cameroon became unstable for some reason. There was an eruption of carbon dioxide gas from deep in the lake, and once an upwelling had started, it was like a bottle of champagne blowing its top. Enough carbon dioxide poured down into the valleys below the lake to smother 1,700 people.

The geophysics behind this sort of eruption is remarkable. The lake is in a volcanic area, and it is about 200 meters deep, with an unremarkable upper 50 meters of ordinary water over a CO_2-rich lower level, where the gas is concentrated but remains dissolved because of the greater water pressure. In the depths of the lake, this saturated water is quite stable, but bring some up to a higher level and a few gas bubbles can start to form. As these rise, some more of the saturated water is carried up and more bubbles form.

As the new bubbles rush to the surface, they push water in front of them, and water rises behind, until the trickle of bubbles turns into a torrent, a huge fountain of gas and gas-soaked water, roaring to the surface, splashing high into the air, and releasing its load of heavy gas, where it forms a spreading blanket that can smother life. The carbon dioxide soon reaches the rock wall that holds the lake in place, and plunges over, an invisible and deadly torrent of gas, sweeping down into the valley below.

After its outburst, the lake settles again. Over time, more and more carbon dioxide would seep into the water from soda springs fed by volcanic fissures deep in the lake, until the whole bottom area was unstable, and the slightest swirl enough to form a few stray bubbles and send the geyser up, creating another murderous cloud capable of killing all before it. But no more. In February 2001, a 14.5 cm diameter poly-ethylene pipe was sunk into the lake, to a depth of about 7 meters from the bottom. This is now being used to bleed off the gas, hopefully keeping it below the unstable point, and plans are in place to add extra pipes.

These pipes will, I hope, be in action soon, as the lake is held in place by a natural dam of loosely packed volcanic rock and waterfalls run over the edge in the wet season. This wall is heavily eroded, and the top 40 meters could fail at any time. This would certainly trigger a catastrophic degassing—as well as flash floods in populated parts of Nigeria, 150 km from the lake.

Tests in 2003 showed the carbon dioxide levels were falling slowly, and a report in *Science* indicated that nearby Lake Monoun was also to be "bled," and engineers were trying to work out if Lake Kivu, between Rwanda and the Demo-cratic Republic of the Congo, could also be drained of its gases—Kivu contains large amounts of flammable methane as well.

Carbon dioxide might be a suffocating agent rather than a poison, but the end result is the same.

Carbon monoxide (CO) is a true poison, and a highly effective one because it is colorless, odorless, tasteless, and non-irritating, so it sneaks up on its victims. Unlike oxygen and carbon dioxide, the CO molecule takes hold of our hemoglobin molecules and will not let go. The stability of the resulting carboxyhemoglobin unit means that no oxygen is carried to the cells. Within a few minutes, our brains are so starved of oxygen that we become brain-dead before we die. Carboxyhemoglobin makes the skin and internal organs go bright red. The absence of this coloration has trapped many a murderer who thought that smoke and flames would cover up the evidence of their crime. A dead body, planted at a fire scene, will not have the bright red blood of a genuine fire victim who has been asphyxiated.

Inhaling CO from car exhausts is a common method for suicide in Australia. Accidental deaths are common in the United States, which uses more fuel-burning heaters. This is not a new problem. A 1935 study in Philadelphia revealed 95 cases of coronary thrombosis during autumn and winter, and only 14 in spring and summer—the difference was attributed to the increased use of CO-producing fuels and poor ventilation in the cold months.

Traffic also generates huge amounts of CO, and toll collectors can be at risk, as can ordinary pedestrians. In 1962, during an oil importation crisis, the only traffic in London was diesel-powered. Diesel produces very little CO, and the measured levels fell almost to zero, except in smokers, who get sublethal doses of the poison from the partial combustion of their tobacco.

CO's effects must have been known before people realized that it was formed when carbon-based fuels burn in low oxygen supplies. While a wood fire dies down comparatively fast, a coal fire is slower, and may exhaust the available oxygen, explaining this reference:

> It was then believed that sea or pit-coal was poisonous when burnt in dwellings, and that it was especially injurious to the human complexion. All sorts of diseases were attributed to its use, and at one time it was even penal to burn it. The Londoners only began to reconcile themselves to the use of coal when the wood within reach of the metropolis had been nearly all burnt up, and no other fuel was to be had.
>
> Samuel Smiles, *Industrial Biography*, 1863

Sea coal was so called because it was carried to London from the mines in the north by ships; while plain "coal" in that period was charcoal. The preponderance of coal fires, however, contributed to London's smog problem, yet another mass killer in its time. The Great Smog of London in December 1952 is credited with killing 4,000 people and led directly to the Clean Air Act that stopped the pea-soupers that had made life so easy for villains like Jack the Ripper.

Toxic gases can harm by mischance, with no ill will involved, but there are those who feel that Union Carbide was taking advantage of the ignorance of those living around the site when they built their factory in Bhopal, India. These unfortunates would become the victims of one of the worst chemical leaks in recent years. The factory made carbaryl, a general-use pesticide, but the actual leak was of an intermediate product, methylisocyanate. Little was known of its toxicity, which did not help in the treatment. As we know now, it is a toxic and corrosive gas, and in December 1984 a large amount of it, probably 30 or 40 tons, escaped into the atmosphere. The poison leaked for three hours before it was shut off, which points to remarkably sloppy management. The number of dead will never be known, but an Indian court's estimates were of 3,000 dead, 30,000 with permanent injuries, 20,000 with temporary injuries, and 150,000 with minor injuries. Other estimates list 7,000 deaths at the time, rising over two decades to a toll of 20,000, with 100,000 victims left chronically ill. As of this writing, the site is still contaminated and toxic materials have leached into the water supply, poisoning drinking water for local residents.

Not only in the United States but around the world, Love Canal is synonymous with "toxic waste dump disaster." There have been other cases in other countries, but few have had the ramifications of Love Canal, which was used as a dump between 1942 and 1952 for 22,000 tons of chemicals that, in the lab, would require labeling as hazards. Mercaptans, phenols, chlorobenzenes, and other nasties were dumped at this site, close to the Niagara River in the city of Niagara Falls, New York.

The Love Canal was a clay-lined hole in the ground that seemed ideally suited for dumping waste, but trouble came after 1953, when development of the 16-acre landfill and its surroundings added a school and 200 homes, which meant digging, and breaching the seals on the toxic waste below. During the 1960s, people noticed unpleasant smells, and toxic chemicals were found in nearby waterways. In response to public outcry over this and other hazardous waste emergencies, Congress passed the Superfund law in 1980. The idea was that polluting companies would be forced to clean up their own sites, but the law also provided for a general fund to be created from targeted industry taxes.

By 2004, some $400 million later, Love Canal was off the Superfund list, with a 40-acre site sealed and capped, and with controls in place to trap any leaching. The toxins will remain until they are released by some future ice age, but by then humans will probably be long gone, and other species will reap our harvest. In the interim, the area has been declared safe, and houses that once had to be evacuated are able to be sold again.

Across America, only about 1,300 of the worst locations are officially listed as Superfund sites, and about 900 of these are regarded as close to fixed. That said, there are probably still 100,000 sites, large and small, across the U.S., where some sort of remediation is required. Many of these will be dealt with through bioremediation, where German fungi consume asbestos, weeds clear lead from soil in Hartford, Connecticut, bacteria in a Wisconsin mine take zinc out of solution, tailored bacteria being developed in Baltimore tear apart noxious PCBs, and brake ferns in Florida strip arsenic from contaminated ground. In others, well, we just don't know what will happen.

All in all, toxic waste dumps don't sound like something you would want near human habitation—yet every city has one, somewhere, and people have to live with them. In 2000, the Sydney Olympics were held on the remains of a toxic waste dump, and while it was no great secret, nobody really cared, because even before Sydney was awarded the Olympics in 1993 plans had been under way to solve the problem. The polluted soil would not be carted away to create a hazard somewhere else: it was bulldozed up into large mounds.

You might be forgiven for asking, "But how would this solve anything?" It works like this. Water seeps down through such hills, leaching out the poisons they contain. So the next step is to put in a drainage system that catches this seepage and returns it to the mound. Then each hill is completely covered in plastic, to stop excess water penetrating the polluted levels, and the plastic is covered in a thick layer of clean soil. The pollutants are thus stabilized and contained, for the present, but at some time in the future, when bioremediation is available, tough microbes will be sent in to shred the poisonous molecules, one by one. Poisoning can be undone, but only at a cost far greater than the profit some slick operator made by dumping the waste in the first place, for which we all pay.

Then again, organisms are sometimes excellent at protecting themselves against poison, using all sorts of biochemical tricks. Evolution sees prey developing ever better poisons, so as to discourage predators. As a rule, the more toxic something is, the more likely it is to survive and breed. This in turn means predators must also evolve, because only those consumers able to withstand or deflect the poisons survive.

Arsenic is the fourteenth most common element in the Earth's crust, and many organisms, including humans, need a defense against arsenic. Accordingly they have developed the detoxifying trick of biomethylation, that is, adding one or more methyl groups (CH_3) to heavy metals and making the metal easier to excrete. In the last half century, we have learned that, in addition to arsenic, selenium, mercury, tin, lead, bismuth, and antimony are also methylated under natural conditions. While this may help an organism to get rid of them, most of these methyl adducts are very toxic when they are disposed of—as we saw in chapter 6, methylation of the arsenic in Schweinfurter green may have killed Napoleon. And it was most certainly the methylation of mercury that did the harm at Minamata, and made the name of a bay a synonym for mercury poisoning.

At Minamata, in Japan, a paper mill released methylated mercury salts that got into the human food chain and harmed many, but even natural background mercury can be a problem.

In the 1960s, Hydro Québec built a number of new dams to generate hydroelectricity, as did other Canadian provinces. Not long after, the local Indians began to show signs of mercury poisoning. These were people who typically ate a lot of fish, so the fish in the dams were tested and found to be overloaded with mercury. This puzzled the authorities, because these dams were in pristine wilderness, not downstream from factories spewing toxic effluent. Only four years after a new dam had filled, most of the fish in it were at or above the Canadian safe limit for mercury (0.5 parts per million), but explaining where it had come from involved some clever geology and physics.

Since the last Ice Age, Canada has acted as a mercury sink, a place where traces of mercury, vaporized in warmer places, settled out, like the condensation of water on a cold bottle. The mercury found its way into the soil and plants, mainly as relatively harmless inorganic salts. When the lakes created by the dams filled, large amounts of mercury went into solution quickly. In the deeper waters, the dead vegetation on the flooded ground made conditions oxygen-free and perfect for mercury methylation, driven by anaerobic microbes, to occur. Once the methyl mercury was in the water, it was concentrated up the food chain, until pike showed levels well in excess of the safety limit, and people eating the pike started to be affected.

The solution was to empty the lakes and refill them, flushing away the mercury, which was, after all, limited. When the lakes refilled, the lake waters returned to safe levels, and so did the fish, but the damage was done so far as people were concerned, because they were well and truly poisoned.

PROFITABLE POISONS

Whatever poison you care to nominate, there is always something worse, and it may be hiding in something as innocuous as a motor vehicle's airbags. These contain sodium azide, NaN_3, and each vehicle contains around 250 grams of the white, salt-like material. To inflate the bag, an electrical heater breaks the sodium azide down to sodium metal and nitrogen gas, but if the sodium azide comes in contact with water, hydrazoic acid, HN_3, is released.

In the ground, low concentrations of sodium azide—as little as 200ppm—can sterilize the soil. In a worst-case scenario, the contents of an average car would produce enough hydrazoic acid to put 5,000 people in a coma, and while it should never come to that, it might. Sooner or later, a car fitted with airbags will go into a wet crusher, or some sodium azide will be spilled into a waterway. In 1996, a truck carrying 50 44-gallon drums of sodium azide overturned and burst into flames 50 miles south of Salt Lake City, Utah. A huge plume of toxic vapors went up, forcing the evacuation of 2,000 people from the small town of Mona, Utah. The same plume in a city would be a much more serious problem. Sodium azide is a disaster waiting to happen.

Around the world, stores of poison are waiting to wreak havoc as people go about their daily business of making a profit. In 1995, 3 million cubic meters of waste water containing cyanide and copper spilled into the Essequibo River from a dam at a mine in Guyana, South America. In 1998, heavy metal pollution from a Spanish mine damaged wildlife in the Guadiamar River and the Doñana National Park. The 1990s also saw cyanide spills in Latvia and Kyrgyzstan, but these were nothing compared with a spill on the night of January 30, 2000, at Baia Mare in Romania.

Cyanide is commonly used to separate gold from tailings left at old mines, but the cyanide sludge is typically left in a dam— and dams fail. The Baia Mare dam did just that, and around 100 million liters of waste water poured northward down the Somes River into Hungary and the Tisza River. The Somes and Tisza were scoured of all life by the torrent of heavy metal– and cyanide-laden water as it started its three-week journey of 1,000 kilometers—a journey that would take it down the Danube all the way to the Black Sea, and leave in its wake a trail of dead fish and less visible but more deadly disruption to the entire ecosystem. Plankton, invertebrates, the 400 protected otters on the Tisza, and the local eagles—all will be affected for years to come.

Cyanide levels in the Tisza River were measured at 12 milligrams per liter, 100 times the level for a "very polluted river," and well above the European Union's safe level for drinking water of 0.05 milligrams per liter. Copper levels were reported at 36 times the "very polluted" level. Over time, the rivers have begun to recover, but Hungarian scientists warn that it is only a matter of time before such a disaster happens again. Most of the water in Hungary's rivers originates in other former Soviet-bloc countries, and their banks are lined with old mines, chemical factories, and worse.

In August 2002, it looked as though they were about to be proven horribly right, as 100-year floods rolled through Europe for the second time in a decade. We have been poisoning our atmosphere in a gentle way with carbon dioxide for hundreds of years, despite Swedish chemist Svante Arrhenius's warnings in the 1890s that this would lead to global warming. Some said that it was still open to doubt, that maybe the thermometers were wrong. But by 2002, few could question that global warming was real, that our climate was changing.

Suddenly, people began to think of what sat on the river banks—all the great rivers of Europe were being flooded with dioxins, mercury, arsenic, lead, and bacteria from sewage, and they in turn were flooding the great cities of Europe. The owners of the Spolana chemical plant at Neratovice, north of Prague, went into overdrive, denying that any of their stored unpleasantness could have or had ended up in the River Elbe, but environmentalists were less sure. Unfortunately, the environmentalists were right, and tons of hazardous chemicals were washed into the mines and released into the air. The effects are still being measured today.

In the nineteenth century, and even into the twentieth, the new industrialists could get away with wholesale pollution; they could dump waste into rivers in an out-of-sight, out-of-mind approach. Dye makers, paper mills, slaughterhouses, foundries— factories of all sorts just pumped the waste into the nearest river through underwater pipes to allow the current to dilute the waste and disappear it. People took this in their stride; but in the 1960s, when people finally began to question the status quo in almost all walks of life, Rachel Carson published *Silent Spring*. Industrialists have been forced, over the past 40 years, to modify their existing practices and initiate new drives toward bio-mimicry rather than biocide.

Once it took many years for a new poison and its risks to be identified, but now, if anything, we are too quick to label something as evil and toxic. There is a middle ground to be found, somewhere between allowing the profit-hungry to destroy this earth and letting the media-hungry destroy our lives and our joy in life. The question, though, is whether that middle ground encompasses the official use of poison.

8

POISONOUS
POLITICS

The man . . . kept his hand upon Socrates, and after a while examined
his feet and legs; then pinched his foot hard and asked him if he felt it.
Socrates said no. Then he did the same to his legs; and moving gradu-
ally up in this way let us see that he was getting cold and numb.
Presently he felt him again and said that when it reached the heart,
Socrates would be gone.

The coldness was spreading about as far as his waist when Socrates
uncovered his face—for he had covered it up—and said (they were his
last words): "Crito, we ought to offer a cock to Asclepius. See to it, and
don't forget."

Plato, *The Last Days of Socrates*, c. 395 BC

Punishment by purging or worse has long been part of a
certain political style. Mussolini's thugs would force political opponents to consume castor oil to cure their waywardness.
Sometimes they would administer as much as a liter of the oil,
enough to cause severe purging, but not death. At other times
the oil was boosted with petrol, in which case it would probably
kill. Of course, poisoning for the good of the state or the

community went back much farther, to sometime before Socrates was condemned to drink hemlock.

It may have been a wise move to accept that fate in order to win immortality. Socrates was something like 71 years old and had experienced a good life. Thanks to his dramatic death, today we recall him and his fate far better than we recall the death of almost any other Greek philosopher.

The state poison, as it was referred to, used a species of hemlock known as *cicuta*, perhaps mixed with other poisons— there is still some debate about this. The dose administered often was not fatal, and the condemned person often needed a top-up. An account of the execution of Phokion in 318 BC describes how "having drunk all the Hemlock juice, the quantity was found insufficient and the executioner refused to prepare more unless he was paid 12 drachmas." Phokion was about two years old when Socrates died, and he lived to more than 80, but he, too, chose to die with dignity.

The name of the poison has been a major source of confusion, because *hemlock* means several different things to us. The hemlock of the Greeks caused numbness that increased until the heart failed or breathing stopped. The symptoms were like those of suffocation, and in *Phaedo*, Plato describes a slowly rising paralysis that started at the feet and moved up the legs to his chest, with Socrates' mind staying clear until the end.

This gentle death described by Plato just did not seem to add up with what modern observers thought they knew about hemlock, but it seems most authors relied on information about the wrong poison. Working on the assumption that he had been given *water* hemlock, they said, rightly, that Socrates' death should have been more violent.

If that were the case, how should we explain Aristophanes' description of death by hemlock in *The Frogs*, written six years before Socrates' death? As an account, it matches that of Plato. The confusion probably rests with Nicander, who may have been a Greek-speaking doctor in the Roman army, because he has given us an account of hemlock poisoning very different from the versions we have from Plato and Aristophanes. In his *Alexipharmaca*, Nicander describes hemlock victims as suffering from a terrible choking and convulsions, symptoms also observed by Swiss anatomist Johannes Wepfer in 1679. Wepfer noted the differences between his firsthand report and the account in *Phaedo*. He took it that water hemlock and poison hemlock

worked the same way, and so felt compelled to question Plato's version.

It was a nineteenth-century Scottish toxicologist, John Hughes Bennett, who got close to the truth, thanks to Duncan Gow and his children. Gow was a poor Edinburgh tailor whose children brought him a parsley sandwich in which the "parsley" was really hemlock. There were no convulsions and no choking but progressive paralysis over the space of about three hours, followed

Conium maculatum

by death—the tailor remained lucid almost to the end. Bennett established the symptoms, performed an autopsy, and had the plant material positively identified. There could be no doubt Gow's gentle death from *Conium maculatum* was like the death Plato described in *Phaedo*. (It should also be pointed out that, tragic though it was, the children's mistake was a fairly common one. Poison hemlock belongs to the Umbelliferae family, other members of which include carrot, parsnip, celery, dill, and parsley.)

So why was Socrates condemned to death? Basically, because he got up people's noses, but the story goes back a few years before the execution. In 410 BC, the Athenian oligarchy was replaced by a new democratic regime. In 406 BC, the Athenian navy won a sea battle against the Spartans, but, with a storm brewing, they headed for home without picking up the sailors who had fallen overboard, fearing the whole fleet would be lost in the storm. Some 2,000 men were lost, and the popular assembly was baying for blood. They wanted the leaders subjected to a mass trial and executed for cowardice for leaving the sailors to drown.

Socrates, president of the assembly that day, declared the mass trials illegal and agreed to separate trials. In doing so, he made some dangerous enemies, and by 399 BC, they were after *his* blood. The vote was 280 to 220 to condemn Socrates on the pretext of "corrupting Athenian youth." As his accuser, Meletus, had called for capital punishment, it was now up to Socrates to propose an alternative, such as exile. Instead, he said he deserved reward as a public benefactor, and while he offered to pay a token fine of one mina, which he later raised to thirty minas, he made no serious counterproposal, and so was sentenced to death.

Phokion, on the other hand, was a general and statesman who openly supported the aristocratic party of Athens, and his sympathies for the traditions of Sparta gave his opponents the leverage they needed. Accused by his political enemies of complicity with the Spartans, Phokion was also sentenced to death, and like Socrates, obeyed the judicial decision by drinking hemlock.

It is a paradox that even today, people who would normally shy away from murder, especially murder by poison, seem more willing to condone murder when it is done in some abstract way, for the good of the state. Right-to-lifers see no contradiction in enthusiastically supporting any move by the authorities to execute murderers by poison, rope, or chair. Lethal injection is now the most common form of execution in the United States.

The same authorities take a dim view of others poisoning.

When Richard Roose allegedly tried to kill Bishop John Fisher and some of his household with arsenic-laden porridge during the reign of Henry VIII, an act made poisoning punishable by being boiled alive. Two others died as Roose had done before the punishment died with Henry.

Poisoning saw a brief upsurge at the end of World War II, with a number of prominent Nazis taking poison tablets rather than face the dishonor of defeat or the War Crimes Tribunal. A few details have now come to light of a vengeful poisoning attempt launched against German prisoners by Holocaust survivors. The German prisoners were members of the Nazi elite guards, awaiting their trial for war crimes. They were being held in Stalag XIII.

In the spring of 1946, a small group of Jews who had escaped from the ghetto in Vilnius, Lithuania, banded together under the name Avenging Israel's Blood. Having discarded their first plan, poisoning a German town's water supply, an act as genocidal as any death camp, they targeted people they saw as more deserving of punishment. They used arsenic (probably arsenious trioxide) smuggled in from France to paint 3,000 loaves of bread bound for Stalag XIII with what they hoped would be a lethal amount of arsenic.

Accounts vary, but probably at least 2,000 prisoners became ill. It is less certain how many, if any, died. Very little has been published on the incident, perhaps because it taps straight into the medieval myth of the Jew as poisoner, a common excuse for anti-Semitism.

POLITICAL POISONING

Certain states seem more at ease with the use of poison than others. There were two earlier attempts on Bulgarian dissident Georgi Markov's life before he was killed in 1978. The first involved a toxin being slipped in his drink, and he was also attacked in Sardinia, before he had a ricin-filled pellet fired into him at close range by a Bulgarian agent using a modified umbrella.

Markov's was the second such attack, as Vladimir Kostov had been similarly attacked in Paris the month before. Kostov became ill but did not die, and his pellet was only found after the fatal pellet was removed from Markov and Kostov was re-examined. Two years later, Boris Korczak, a CIA double agent and former Soviet citizen, was also shot with a ricin pellet, and while some reports say he died, he actually survived.

In most cases, it is not enough for a poison to get into the body: it must enter the cell itself, and this means getting past that amazing surface, the cell membrane. Much of the membrane is a sheet of fairly inert stuff called phospholipids, but across the surface there are dotted some special carriers that take certain kinds of molecules as they drift past, and either admit them because they have the right key, or actively gather them in.

Most chemistry involves things spreading out to a randomly even distribution, but the active carriers in the membranes buck this tendency (as long as some poison doesn't come along and block the carrier!). There are, however, other ways in which a chemical, even a poison, can end up inside a cell.

A molecule or particle may be engulfed, taken into a pocket of the cell membrane that wraps around the particle and then opens up on the inside. This is how insoluble uranium dioxide and asbestos get into the cells of the lungs. Then there is *facilitated diffusion*, which requires a membrane carrier but also needs a diffusion gradient, *passive diffusion* through the phospholipids of the membrane, and *filtration* through pores.

Perhaps the most fearful of poisons in recent years, at least in the view of the media, is ricin, a small and rather nasty molecule from *Ricinus communis*, the castor oil plant. The plant is common in many parts of the world, the toxic material is fairly easy to separate out from the seed, and it has a particularly diabolical way of disrupting the cell, so just one molecule could kill a cell. In addition, ricin needs no carrier.

Ricin is a smallish protein in two parts, called A and B. The A chain is 267 amino acids long, and globular. The B chain is 262 amino acids long and shaped like a barbell. The chains are linked by disulfide bridges, and the B chain hooks onto the outside of a mammalian cell by binding to the galactose part of a glycoprotein. (Translation: it finds a suitable hitching post on the outside of a cell and locks on.)

The cell membrane then forms a small bubble, a vacuole that moves into the cell, taking the ricin inside. As the vacuole re-forms inside the cell, the A and B chains separate. The B chain makes a channel through the vacuole membrane, letting the A chain enter the cytoplasm, where it gets to the ribosomes and blocks protein synthesis. The A chain on its own is an enzyme able to remove a specific component from ribosomal RNA. In this way it inactivates the ribosome and so stops protein synthesis. Because it acts as a catalyst, the single subunit is not used up, and it can inactivate many ribosomes, about 1,500 a minute. While there can be a few million ribosomes in a cell, the damage soon begins to take effect, as the cell goes out of control.

The disulfide bridge linking the two halves of the ricin molecule is a simple break point that is intended to be fractured when the time is right. An enzyme is able to carry out its function because it is the right size and shape: in fact it only stays in the right shape because there are disulfide bridges holding the protein in a particular arrangement. This is a weak point which can be targeted by heavy metals such as thallium and arsenic.

In 2004, Viktor Yushchenko, who won the presidency of Ukraine in early 2005, was allegedly poisoned by someone while campaigning. His symptoms were dismissed by opponents as the result of eating "bad sushi and cognac," but his face showed clear signs of chloracne, a symptom of dioxin poisoning. Tests in Vienna in December 2004 pointed to dioxin as the cause. As yet, nobody has been accused of poisoning him, but there are several suspects.

Political assassination by poison is almost as old as history. Generations of the Borgias turned the Vatican and Pontificate into a scandal. According to some, the family used a white powder, *La Cantarella*, that may have contained arsenic, or maybe phosphorus, or perhaps lead acetate, but most probably all three. It was supposed to have a sugary and pleasing taste and could be added to food or drink. The family kept a staff of Italian astrologer-chemists who worked with mercury, arsenic, phosphorus, hemlock, monkshood, henbane, yew, and poppy. Some of the activities attributed to the Borgias may have been exaggerated. All the same, they gave the writers of their day plenty of material:

> Can there be a more express act of justice than this? The Duke of Valentinois having resolved to poison Adrian, Cardinal of Corneto, with whom Pope Alexander VI, his father and himself, were to sup in the Vatican, he sent before a bottle of poisoned wine, and withal, strict order to the butler to keep it very safe. The pope being come before his son, and calling for drink, the butler supposing this wine had not been so strictly recommended to his care, but only upon the account of its excellency, presented it forthwith to the pope, and the duke himself coming in presently after, and being confident they had not meddled with his bottle, took also his cup; so that the father died immediately upon the spot, and the son, after having been long tormented with sickness, was reserved to another and a worse fortune.
>
> Michel de Montaigne, *Essays*, 1575

The Duke of Valentinois was Cesare Borgia (1476–1507), who became the bishop of Pamplona at 16 and a cardinal at 18, while his father became pope. There are variations between versions, but the two did indeed dine with Corneto. The next day the two Borgias and some other guests were seriously ill. The cause may have been malaria, but according to rumor the wine had been poisoned. Alexander died rather horribly, but Cesare was a fit young man, and though critically ill for weeks, he survived. The other popular rumor at the time was that the dishes had been switched, and the wrong persons fell victim.

Cesare was reputed to have a ring that had twin lion heads with sharp teeth. He would greet victims with a warm hand-shake, and that was that. His father, Pope Alexander VI, had a cabinet in his Vatican apartments that he would ask victims to unlock for him. The lock was stiff and there was a sharp point in the handle of the key that pricked the palm of the user, who would soon be dead. Lucrezia Borgia, on the other hand, was a famous beauty who took three husbands, but there is no evidence that she followed the family poisoning tradition.

Physostigma venenosum

Political assassination does not have to be done by stealth. Cultures around the world have a history of legalizing the removal of unwanted citizens by sentencing them to trial by ordeal. Unfortunates could be tortured to within an inch of their lives, and beyond, or forced to ingest a lethal dose of their inquisitioner's poison of choice. The guilty would succumb to the poison, the innocent, theoretically, vomit and survive. One poison used in this way is the "ordeal" bean or Calabar bean.

To botanists, this is *Physostigma venenosum*, a climbing peren-
nial in the pea family, which can grow to 50 feet in the southeast
corner of today's Nigeria. According to Taylor, the bean was used
in the nineteenth century on the western coast of Africa in an
ordeal for people suspected of witchcraft or accused of serious
crime. The bean, known locally as *eséré*, was also used in duels,
with two adversaries each consuming half a bean. The accused
would be given a drink of the pounded beans infused in water,
which was said to be fatal within an hour. People were so confi-
dent in the accuracy of the test that suspects would voluntarily
take an emulsion of the seed to prove their innocence.

Taylor reports Sir Robert Christison, one of the most famous
toxicologists of his day, as saying that many an innocent person
thus paid the penalty for an undue reliance on superstition—and
Christison would know. In his usual adventurous style, in February 1855 he tried one-eighth of a bean, six grains, and reported
the results. When he detected no sensation, he doubled the dose
to a quarter of a bean, and reported feeling the will but not the
power to vomit, a slowing of the heart, and drowsiness. This
dose was probably getting perilously close to the safe limit.

It was a simple and direct course of justice and, as such,
heartily disapproved of by the British colonizers. Hoping to end
the practice, they banned the cultivation of the bean, but, given
that the bean flourished in the wild, the ban was not terribly
effective.

Some Western colonists who were prepared to learn from the
local inhabitants discovered that the beans could be swallowed
whole with little ill effect; useful knowledge for anybody captured by hostile tribesmen.

The lack of pain reported by Christison led some to suggest
that the Calabar bean might be used to dispatch criminals, but
before long other, more therapeutic, properties of its active

ingredient, physostigmine, were being noticed. Where atropine (belladonna) dilates the pupils of the eye, physostigmine makes them contract, and ophthalmologists greeted the new substance enthusiastically.

Today, new uses for physostigmine are still being found. As far back as 1898, it was discovered that injections of physostigmine caused ejaculation, allowing impotent men the chance of becoming fathers. Paralyzed men have successfully impregnated their partners thanks to physostigmine injections and artificial insemination. Physostigmine is routinely used to counter the effects of curare and atropine. (Appropriately enough, atropine is used to treat those poisoned by the bean!) Physostigmine binds to the same enzyme as sarin, but only temporarily, whereas sarin binds permanently, so there is every prospect that a suitably timed injection could protect those in a battle zone from the effects of sarin gas.

The Pythagoreans abstained from eating beans. While this information has traditionally given rise to intestinal humor in generations of schoolchildren, it might be worth considering that a number of early scholars were fairly peripatetic. The bean family has a number of deadly members, so perhaps there was more common sense to this rule than meets the eye.

It's a short step from sanctioning the use of poisons to remove individuals from the political arena to wholesale removal of opponents from the field of battle.

Poison has had a long, if less than honorable, history, and the scientific advances of the last 50 years mean that ever more insidious and deadly poisons are there for the dangerous, unscrupulous, or downright mad to unleash upon their fellows. Such is humanity.

9

POISON
AND WAR

"Peace on earth!" was said. We sing it,
And pay a million priests to bring it.
After two thousand years of mass
We've got as far as poison gas.

Thomas Hardy, Christmas: 1924

Two millennia before Hardy penned these bitter words, Spartan allies during the Peloponnesian War captured an Athenian-held fort by piping smoke from lighted coals, sulfur, and pitch into the fort through a hollowed-out beam.

This might be the first documented use of poison gas in war, a stratagem which came to the fore in the nineteenth century. In 1812, the British Admiralty had decided that the use of burning ships loaded with sulfur before marine landings in France would be against the rules of warfare, but such scruples would be long forgotten in only 100 years.

In 1854, as Europe languished in the grip of the Crimean War, Sir Lyon Playfair suggested the use of shells filled with cacodyl cyanide against the Russians to break the siege of

Sebastopol. The War Office called this "as inhumane and as bad as poisoning the enemy's water supply," suggesting they had not read their Pausanias. Playfair answered that it was considered legitimate to spray the enemy with molten metal that produced the "most frightful modes of death," and so he found the response incomprehensible. Perceptively, he commented that "no doubt in time chemistry will be used to lessen the sufferings of combatants."

Poison in war was an acceptable stratagem in ancient times. Pausanias tells how Solon defeated the people of Cirrha by diverting a river that flowed through a canal into the besieged town. When the people of Cirrha still held out, drinking rain and well water, Solon had hellebore roots thrown into the stream, and then allowed the waters to flow again. The Cirrheans all drank deep of the water; the sentries were struck down by diarrhea, and left the walls, allowing the town to be taken.

In 1864, a proposal for the Union army to use chlorine shells against the Confederacy was also rejected, but the idea was clearly alive and well. The 1874 Brussels Convention tried to preempt the inevitable by banning the use of poisons in war, while the Hague Conventions of 1899 and 1907 only offered a weak and vaguely worded resolution against the use of chemicals on the battlefield. Time was slipping away while the diplomats fiddled, and science was advancing as World War I approached.

Of course, much of the science had been in place for a long while. Phosgene and chlorine, two of the most deadly gases, had been around for more than a century when war broke out in 1914. Scheele had first prepared chlorine in 1774, and it was named in 1810 by Humphry Davy, who proved it to be an element. Two years later, his little-known brother John discovered phosgene or carbonyl chloride, $COCl_2$, when he exposed carbon monoxide and chlorine to sunlight while they were contained in a glass vessel. John Davy described it as "producing a rapid flow of tears and occasioning painful sensations."

By 1914, even more chemical weapons were available. In Paris, gendarmes had been using unpleasant gases to control rioters for a number of years, and it seems the first chemical shots in the war were actually French antiriot agents lobbed into German lines, although it is more usual to accuse the Germans of being the first to use poisonous gases to kill. History may not always be written by the victors, but they do tend to get the major blame-allocation rights.

Poison gas, in the strict sense, was first used on a battlefield in April 1915. The German plans of moving across the country fast and outmaneuvering their enemy had become bogged down in trenches, barbed wire, and mud. Infantry sat in their trenches while the artillery behind them lobbed high-explosive shells into the enemy's trenches. Once the enemy was sufficiently softened up (they hoped), machine-gun fodder would be sent forward to a point where they could be mown down by the surviving defenders.

In spite of all the snide comments made about military intelligence, field commanders had no problem working out that a barrage from the other side indicated an impending attack, so reinforcements would be readied and the frontline troops would shelter in bunkers, ready to emerge and man the machine guns once the barrage stopped and the enemy advanced.

It was a total stalemate, and it was likely to go on until there was one soldier left standing. The lack of progress came down to access to resources: the Allies could milk their overseas colonies of both fuel for what was increasingly a mechanized war and nitrogen compounds for explosives. Given time, they should have been able to wear Germany down. Germany's biggest challenge was maintaining the supply of explosives, and this need had been met largely by one man, Fritz Haber.

Nitrogen is available for free—it makes up about 80 percent of the atmosphere—but the stuff in the atmosphere is almost inert, and almost useless for fertilizing plants or making useful compounds. Nitrates were needed to fertilize crops, and guano supplies in Chile were running out. This was the situation confronted by Sir William Crookes in a presidential address to the British Association in 1898. More wheat was needed, he said: "The world's demand for wheat—the leading bread-stuff—increases in a crescendo ratio, year by year."

Haber took up the challenge, and developed a method of forcing nitrogen to react with hydrogen to make ammonia, the first key step in synthesizing nitrates. Haber's method proved to be highly successful—the first plant opened in 1913, and even now the Haber process is still used in 600 plants, which generate a world total of more than 100 million tons of ammonia each year.

It is brutal chemistry, combining the gases at a temperature of 400–500°C and a pressure of 20 atmospheres or more in the presence of an iron catalyst. Curiously, the ammonia is a poison to us, but the catalyst is said by chemists to be "poisoned" by oxygen. There is just no way we can escape poisons, it seems, but a third of the world depends on Haber process products for its continued survival.

As the war in the trenches lost momentum, Haber's plant quickly became crucial to the German war effort. Perhaps Haber can be accused of contributing to the stalemate because his process maintained the supply of ammonia needed to make nitrates and nitric acid for the German munitions industry.

Apparently Haber commented to Gustav Krupp von Bohlen und Handbach that, after the first few months in the trenches, each side had become better at defense than offense. A new weapon was needed, he said, to break the murderous stalemate of trench warfare. Echoing Lyon Playfair, Haber actually saw the use of gas as humanitarian, because the gas immobilized, rather than killed, its victims. There has to be some doubt about the German claim that gas warfare was only inspired by the stalemate, and certainly the victorious Allies would later counterclaim that the Germans had planned their gas attacks long before the war began. For whatever reason, on April 22, 1915, the Germans launched their third attempt to break through and take Calais. They were tired of mounting mass attacks and getting slaughtered under machine-gun and rifle fire. For the first few days, the nature of the gas was unknown, and it was referred to only as "asphyxiating gases." We know now it was 500 tons of chlorine, released from 200,000 cylinders, and the Allies' bewilderment seems a little strange. Any worker from St. Helens or Widnes could have told them what it was: the Germans were attacking them with Roger.

The gas was identified in the first reports as "contrary to the rules of The Hague Convention," a rising cloud of greenish-gray iridescent vapor. The Germans, said reports, were prepared to work within the gas, as though this somehow made it more reprehensible. Men dressed in what looked like diving suits could be seen handling the cylinders, with hoses directed at the French lines. About 800 Allied defenders were killed, while a further 15,000 were forced to flee.

According to the *New York Tribune*, the German troops following up "held inspirators in their mouths to protect them from the fumes":

This new form of attack needs for success a favorable wind. Twice in the day that followed the Germans tried trench vapor on the Canadians. . . . In both cases the wind was not favorable, and the Canadians managed to stick through it. The noxious, explosive bombs were, however, used continually against the Canadian forces and caused some losses.

New York Tribune, 1915

Victor LeFebure recorded General Sir John French's original, horrified reaction:

Following a heavy bombardment, the enemy attacked the French Division at about 5 p.m., using asphyxiating gases for the first time. Aircraft reported that at about 5 p.m. thick yellow smoke had been seen issuing from the German trenches between Langemarck and Bixschoote. What follows almost defies description. The effect of these poisonous gases was so virulent as to render the whole of the line held by the French Division mentioned above practically incapable of any action at all. It was at first impossible for any one to realise what had actually happened. The smoke and fumes hid everything from sight, and hundreds of men were thrown into a comatose or dying condition, and within an hour the whole position had to be abandoned, together with about fifty guns. I wish particularly to repudiate any idea of attaching the least blame to the French Division for this unfortunate incident.

Victor LeFebure, *The Riddle of the Rhine*, 1923

Still, the Germans failed to break through, and the Allied troops were given temporary cotton masks, which they had to soak in their own urine. The urine broke down to release ammonia (itself a poison under the right conditions), which neutralized the chlorine. In July 1915, they got their first "efficient" gas masks and respirators, although Robert Graves was rather scathing about their prophylactic qualities:

This, the first respirator issued in France, was a gauze-pad filled with
chemically treated cotton waste, for tying across the mouth and nose.
Reputedly it could not keep out the German gas, which had been used
at Ypres against the Canadian Division; but we never put it to the test.
A week or two later came the 'smoke-helmet', a greasy grey-felt bag
with a talc window to look through, and no mouthpiece, certainly
ineffective against gas. The talc was always cracking, and visible leaks
showed at the stitches joining it to the helmet.

Robert Graves, *Goodbye to All That*, 1929

For all that, the Allies had been discussing the possible use of
poison gas since at least 1812, and despite depicting the vile
Hun as a baby-bayoneting monster, the Allies seem not to have
anticipated that someone else might have had the same idea.
General French complained on June 15, 1915, that "All the scientific resources of Germany have apparently been brought into
play to produce a gas of so virulent and poisonous a nature that
any human being brought into contact with it is first paralyzed
and then meets with a lingering and agonising death."

By September 25, the British were ready to retaliate, but
their first trial failed when the wind blew the chlorine back on
their own advancing troops. Soon after, gas began to be delivered
in shells, and the cumbersome arrangements of gas tanks and
hoses could be disposed of.

The problem with chlorine as a weapon was that the victims
started coughing as soon as the gas arrived, making it hard for
them to inhale a lot of it. The coughing reflex also prompted
troops to don their gas masks. Phosgene, a compound of chlorine
and carbon monoxide, was brought into play in December 1915
and proved more effective. It was often used along with chlorine
in a mix called "white star." Phosgene causes at most a minor
irritation of the lungs and throat, and there is no respiratory

reflex, no coughing to protect the victim, so the gas can move deep into the lungs, where hydrogen chloride is released and causes congestion and fluid buildup. The odor threshold for the gas is 1.5 milligrams per cubic meter, and it irritates the mucous membranes at 4 mg/m^3. The lethal concentration/exposure

The French tried a different tack, firing off a total of 4,000 tons of cyanide, as 0.5- or 1-kilogram payloads, apparently without killing a single German, as the gas dispersed too fast to have an effect. Even so, tabloid scaremongers still bleat about terrorists developing "the capability to make cyanide," without considering the challenge of transporting the cyanide and then releasing it effectively. Cyanide gas must be relegated to the role of effective suicide tool and potential indoor killer; it is of little value in the open.

Oddly enough, a few Germans did die of cyanide poisoning during World War I, but only after using the gas to destroy vermin. In Essen, some Krupp workmen's barracks were treated with cyanide gas and improperly aired. Five of the workmen who entered the barracks became comatose but revived, while another ten died from the fumes. In another case, 100 soldiers put deloused clothes on too soon after they had been treated with cyanide and presumably absorbed the poison through their skin: ten lost consciousness, but in the end, none died.

measure (LCt$_{50}$) of phosgene is about 3200 mg min/m^3. At low levels, then, it will affect the victims slowly, and, because it smells like new-mown hay, it has every chance of getting past a soldier's sensory defenses.

By 1917, however, gas warfare had taken a nasty new turn with mustard gas being brought into the war on both sides.

Dichloroethyl sulfide gets its common name because it smells a bit like mustard, though its effects are rather less appetizing. Mustard gas causes irritation, and then turns the lungs solid. Fewer than 5 percent of casualties who were treated died of mustard, but they were typically convalescent for 6 weeks.

By the end of the war, Britain had 2,000 dead and 125,000 incapacitated and hospitalized by mustard, proving its effectiveness against an enemy's war effort. Like chlorine, mustard was first used at Ypres, but on July 12, 1917, it was delivered by artillery shells, and one attack caused 20,000 casualties.

Mustard is a persistent liquid, so it made anything a soldier touched a potential enemy weapon. Most importantly, there was a latent period of several hours, so there was no immediate sign that the victim had been exposed. Horses were still used for transport in 1917, and they also needed protection from sulfur mustard, while men had to wear hot and bulky protective suits.

The substance was first made in the early 1800s, and is sometimes referred to as Yprite or HS, short for *Hun stoffe*. More recently, it was used by Mussolini's army in what was then Abyssinia, mainly to interdict areas, and by Iraq against Iranian troops and, later, against its own Kurdish citizens. British forces used it near Baghdad in about 1920, when liberated Mesopotamians decided they were insufficiently liberated. Britain is also said to have used mustard in Russia in about 1920, and in Afghanistan soon after World War I, while the Spanish used gas against the Riff tribes in Morocco. The Japanese used it in China before World War II was declared.

No one could doubt the power of poison gas as a weapon after World War I. In America, General Pershing wrote, "The effect is so deadly to the unprepared that we can never afford to neglect the question." Yet during the late 1920s and 1930s, while the world's nations were developing new and better armaments, there seems to have been remarkably little work on new poison gases except in Germany, where they arose out of insecticide development.

By 1936, Gerhard Schrader and his colleagues had identified chemical agents that blocked cholinesterase, causing loss of control over respiration and other functions and leading to asphyxiation. The first was tabun (dimethylphosphoramido-cyanidate) and the second was sarin (isopropylmethylphosphoro-fluoridate), named after its developers, Schrader, Ambros, Rüdiger, and van der Linde. Sarin is similar to tabun, but even more toxic—it was the gas used in Tokyo in 1995 by Aum Shinrikyo.

These agents were weaponized and stored, so by the end of 1944 Germany had a stockpile of 12,000 tons of tabun. No one really knows why they weren't used—possibly it came down to fear of retaliation, but Britain had nothing more deadly than a new version of mustard gas. One interesting theory is that Hitler, himself a mustard gas victim in World War I, may have been against the weapons—his senior staff would also have been junior officers in World War I and so equally exposed and inclined to be opposed to the use of gas.

All the same, the British expected gas attacks to come, and the fourth edition of *A Catechism of Air Raid Precautions*, published in 1939, lists ten gases, classified as: persistent and non-persistent; choking (phosgene and chlorine); and tear, nose, and blister (mustard and Lewisite—developed by Dr. Wilford Lee Lewis, perhaps the only human to have a poison named after him, if you discount sarin's inventors). Citizens were expected to read, mark, and inwardly digest that Lewisite is chloro-vinyl-dichloro-arsine; that one of the tear gases, SK, was named for South Kensington, where it was first developed; and that BBC had nothing to do with the British national broadcaster but was in fact bromo benzyl cyanide. If the citizens were going to die of noxious gases, they would die educated.

The details show a strange matter-of-factness. Americans knew Lewisite as Dew of Death. The catechism tells us that it smells pungently of geraniums, although the purer the gas, the less pungent the odor. It says that phosgene smells of musty (rather than new-mown) hay and causes coughing for a few minutes, but that the coughing then goes away for some hours before returning.

One question in the catechism asks why it is necessary to put on and take off protective clothing by numbers. The answer:

> For the same reason that Lewis gunners were taught to deal with stoppages by numbers, i.e., because the process must be done accurately and thoroughly, and usually will need to be done in conditions of hurry and perhaps some excitement. It is found by experience that the best preparation for such hurried work is constant practice under drill conditions, so that in an emergency the act is done rapidly, automatically and properly. The purpose of discipline is efficiency in action. A badly-put-on kit is dangerous by reason of the false confidence it engenders, and the unsuspected risks it exposes one to.

> A Catechism of Air Raid Precautions, 1939

The handy, pocket-sized booklet lists instructions for decontamination and gas-proofing windows with putty and sticky tape; stuffing chimneys and sealing doorways; formulas for converting floor area to breathing-time in a sealed room; and drill procedures, decontamination depot design, and much more, all in

140 pages. The Home Office also produced pocket books on gas and gas defenses, as early as 1937.

This chapter was first sketched out in a time of war, a time when the excitable media were rattling on about stockpiles of cyanide shells and bioterrorists using cyanide in public spaces. In 1937, people seem to have been a little better informed:

> In closed spaces [hydrocyanic acid] is extremely toxic; in the open, however, the dispersion of the gas is so rapid that relatively low concentrations result that are not lethal. This fact explains the failure of hydrocyanic acid gas shells in the last war in the open field, where they caused but few casualties.
>
> Home Office, *Medical Treatment of Gas Casualties*, 1937

In 2000 it came out that members of the chemistry department at Cambridge University tested nerve gases and other chemical warfare agents on themselves during secret World War II experiments. *Nature* quoted chemist Fred Pattison, who went blind for ten days when his dosage was too high. "I thought I was permanently blinded," he recalled. Pattison said everyone accepted the risks at the time, because they saw their work as part of the war effort and as something of national importance. One of the chemists involved in the Cambridge experiments said he shuddered to think what an ethics committee would say today about the research. But as another said, "It was war time. Everyone was taking risks. You were not offering yourself as sacrifice, but it took away the boredom of the war years in Cambridge."

Perhaps there was more to it than that, because this sort of risk-taking seems to be something of a Cambridge tradition. J. S. Haldane rushed off to France for the postmortem of a victim of the first gas attack in World War I, a young Canadian officer killed by chlorine. On his return, his attic became a sort of gas chamber, where he and his son, J. B. S. Haldane, tested assorted toxic gases and gas masks of their own design. Haldane senior's lungs were permanently damaged, but the household carried on blithely. His daughter Naomi (later the writer Naomi Mitchison) and their lodger Aldous Huxley shredded stockings, vests, Naomi's knitted cap, and Aldous's scarf to provide the absorbent filling for the prototype respirators.

In a letter to his father in June 1915, Huxley recounted how he "walked into some nitric acid, which one of Dr. Haldane's assistants had put outside the lab and left in the pathway. It squirted over my foot and leg," and while he did not notice it at the time, some 45 minutes later he suspected a fly bite and, later still, found a brilliant yellow stain on his heel and blisters. So it wasn't just the Haldanes who bore the risks in their house. Huxley did report a small advantage a week later: the limp secured him a seat on a bus, he said.

Poison would be used in World War II in two main ways: in the gas chambers of the Nazi death camps, where cyanide and carbon monoxide took human lives; and in the form of DDT, which saved lives. There were, however, some deaths from mustard gas. In 1943, German aircraft bombed the *John Harvey*, a cargo ship moored in Bari Harbor, Italy. This ship carried 2,000 American 100-pound mustard bombs, and the resulting gas cloud caused 600 military casualties plus an unknown number of civilian casualties. There was reportedly a 14 percent fatality rate, mostly sailors who dived into the mustard-contaminated waters. They swallowed some and absorbed more through the skin, but they were the only combatants to die of recognized war poisons during the conflict.

Zyklon B, originally a pesticide, was the major poison killer of World War II, used by the Nazis to kill millions of Jews and other "undesirables." There were a number of variants of Zyklon, but it was basically hydrocyanic acid on a carrier substance. As we have seen, this is ineffective in open spaces, but in the death chambers of the camps, it was very effective indeed, though it would have taken victims several terror-filled minutes to die.

The only seeming streak of humanity in the entire operation was the removal of the smell agent from the Zyklon B, which served to warn those using it to fumigate for pests of a leak. It appears, however, that this was an economy rather than an attempt to ease the victims' pain. At Treblinka, a simpler method was used. Two engines from captured Soviet tanks were rigged up so that carbon monoxide poured into the death chamber, taking from 30 to 40 minutes to kill the unfortunates herded inside.

There are those who have denied that the horrors of the Holocaust happened, but contemporary evidence and reports from people who entered the camps during the war cannot be denied. There was simply no time to fabricate the evidence, though some still insist that the deaths in the camps were caused by typhus. It is true that typhus carried off many in the camps, but poison did for many more.

Oddly, typhus was the target of the third but less lethal use of poison in World War II, which, like the *John Harvey* incident, happened in Italy in 1943. A typhus outbreak among refugees in Naples led to 1.3 million people being dusted at the rate of 10 pounds to 150 people with a white powder that destroyed the lice that spread typhus. The lice were stopped in their tracks, no humans were harmed, and the powder was greeted as nothing less than miraculous. Winston Churchill said in September 1944, "The excellent DDT powder, which has been fully experimented with and found to yield astonishing results, will henceforth be used on a great scale by the British forces in Burma and by the American and Australian forces in the Pacific and India in all theatres."

DDT is dichlorodiphenyltrichloroethane, which may be more easily read as dichloro-diphenyl-trichloro-ethane (real chemists don't use hyphens, but I find it easier to understand that way). It was first made in 1874, but it was only in 1939 that Swiss chemist Paul Müller recognized its insecticidal properties. The toxic dose of DDT in humans is known to be greater than 10 milligrams per kilogram of body weight, and no human fatalities have ever been recorded. In one study, human volunteers took 35 milligrams per day for a year, with no demonstrable toxicity. Nobody knows to this day how it affects insects, but it probably acts on their motor nerve fibers or the motor cortex, changing the transport of sodium and potassium ions, the key process that makes up a nerve impulse.

Müller was awarded the 1948 Nobel Prize for Medicine or Physiology for his discovery of DDT, but before long the environmental damage that DDT causes started to become apparent. DDT is chemically stable, insoluble in water, soluble in fat, and some of it breaks down to DDE, dichlorodiphenyldichloroethylene. The real problem is in the fat solubility, because when one animal eats another, it retains its meal's DDT and DDE.

The facts and figures are alarming. In one area of California, plankton had 4 parts per million of DDT, while bass in the same area had 138 ppm, and grebes feeding on the bass had 1500 ppm. For some reason, toxic concentrations affect birds and fish, especially in egg production. Human mothers exposed to 0.0005 mg/kg/day produced milk with 0.08 ppm DDT, so the infants were exposed to 0.0112 mg/kg/day/—a magnification of more than twenty-fold.

This has led to a murderous overreaction, with green groups demanding a total ban on DDT, but DDT is used in more than one way. Most DDT is used in wild and uncontrolled agricultural spraying, and this should be stopped as soon as possible. A far smaller amount is used in malaria spraying to kill mosquitoes, and so long as there is no other way to control malaria, this may need to be continued, even though it is undesirable. On the other hand, there is a minor application that loses almost no DDT to the environment: DDT-impregnated pads for mosquitoes to rest and die on.

When the DDT pads are used, if the dead mosquitoes are eaten by scavengers, some of the DDT will enter the food chain, but it is a very small amount compared with the amount spread all over the place by spraying, and it is essential that these pads continue to be made and used, perhaps with more creative management of the distribution and disposal of the pads—such as getting funding for a pad replacement scheme, where old pads are handed in and replaced with new ones. The problem is not DDT; the problem is the way it has been used.

When the POPs (Persistent Organic Pollutants) Treaty was being argued in 2000, Roger Bates of Africa Fighting Malaria, a loose coalition of DDT supporters in South Africa, said banning DDT now would be like "crossing a street with heavy traffic to avoid a crack in the pavement." His group gained support from the World Health Organization, and in the end the treaty required registration of DDT use but did not ban it. This left the way open for Western aid organizations to bring pressure to bear on countries still using DDT, poisoning good science. Any such ban will almost certainly rule out DDT-soaked pads as well.

DDT may have been used successfully to save human life during World War II, but the same cannot be said of the other environmental poisons of the twentieth century—a list that includes PCBs, dioxins, furans, aldrin, dieldrin, endrin, chlordane, hexachlorobenzene, mirex, toxaphene, and heptachlor.

Two of these stand out as needing special discussion: the PCBs and the dioxins. Polychlorinated biphenyls (PCBs) are chlorinated aromatic hydrocarbons that were used in capacitors, transformers, plastics, and in other ways for half a century. There are more than 200 combinations, and most are made as mixtures. Generally, the more chlorine it contains, the more toxic the PCB will be. PCBs are everywhere, even in (or on) Arctic ice. For the most part, PCBs are bioaccumulated in seafoods, although there was a Japanese case where Yusho, or rice oil disease, was caused by rice oil contaminated with PCBs. More than 50 percent of the women affected gave birth to children with abnormalities, suggesting the PCBs may be fetotoxic. Recent studies suggest that at least part of PCBs' toxicity may be due to dibenzofurans, relatives of the dioxins.

Dioxins are truly terrible. The devastation caused by Agent Orange may in fact be due largely to dioxin impurities that were not eliminated during the production of its main components, 2,4-D and 2,4,5-T. American forces used Agent Orange in Indochina during the Vietnam War, mainly to clear the jungle and deny cover to Viet Cong soldiers. This poisoned the trees, broke down the natural regeneration cycle of the rain forest, and triggered major ecological devastation, as well as leaving residues that affected not only the Vietnamese and the American forces on the ground but also the children born long after the war to all those touched by the sprays or by spillages.

In 1966, the United States argued that the Geneva Protocol of 1925, which banned poison gas and germ warfare, did not cover the use of riot control agents or defoliants, and this justified their continued use of the defoliants. In fairness, the prevalent scientific view at that time was that 2,4-D and 2,4,5-T were mere plant hormones and harmless to humans. How wrong we were!

Any dioxin poisoning in Vietnam was accidental and unintended, but it would probably help to clarify just what a dioxin is. Strictly, they are chlorinated dibenzodioxins, two benzene rings linked by two oxygen atoms and able to add up to eight chlorines. As more chlorines are added, there are more and more possible ways of arranging the atoms, up to four chlorines, and then the numbers fall away again. There are two monochloro or octachloro, ten di or hepta, 14 tri and penta, and 22 tetrachloro dibenzodioxins, one of which, referred to as 2,3,7,8 TCDD, is perhaps the nastiest.

Dioxins were shown to be hazardous to at least some mammals in the early 1970s, when many horses in Times Beach, Missouri, were poisoned because dioxin-contaminated oil was used to settle the dust in an exercise yard. Then, in 1976, somewhere between 2 and 7 kilograms of 2,3,7,8 TCDD escaped from a chemical plant in Seveso, Italy. This was when toxicologists discovered an interesting fact they could never have learned from experiments: humans are less affected than most other mammals by 2,3,7,8 TCDD.

After the escape at Seveso, rabbits, chickens, and wild birds died first, then larger animals, but no humans. Some women in the first trimester of pregnancy had terminations, but many did not, and there seems to be no evidence of any more abnormalities in the resultant children than you would expect to find in a random sample.

Biological and chemical weapons excite the righteous wrath of the world's media, and we read time and time again of suspected terrorists with stocks of suspicious white powder, the favorite terror agent since the 2001 anthrax cases in the United States, where weaponized anthrax spores were sent through the mail to a number of outlets. The spores had been mixed with bentonite, and while U.S. law enforcement agents believe they know who was responsible, nobody has been charged, and the media's initial reaction was that this was a typical Iraqi practice.

Such a claim conveniently overlooked the fact that bentonite is a clay mineral that can be bought by the barrel or even the container-load in Texas and Oklahoma, where it is mainly used to seal off old wells and stop surface water from contaminating the water table. Bentonite is also used in some medicines and in kitty-litter—it is not hard to obtain. Second, the strain of anthrax used was the Ames strain, a virulent form commonly found in American biowarfare labs and which also originated in Texas, but that is neither here nor there. Meanwhile, the white powder, which has assumed mythic importance, is not particularly Iraqi at all: it is merely a carrier that is effective in carrying anthrax spores up into the air, where they may be breathed in.

Anthrax occurs in many places around the world, and workers in a number of industries are likely to be inoculated already against the disease, which can, in any case, be counteracted with the antibiotic Cipro. A committed terrorist would have no problem preparing the anthrax white powder, even without a high-quality laboratory, but it is unlikely any anthrax attack would ever rate among the world's great poisonings. Like the French cyanide shells, it would be one of the great poison flops instead.

The same applies to ricin, botulinum toxin, cyanide, and most of the other potential terror poisons as well, if they are sprayed. To be effective, poisons have to be limited to random attacks, a few here and a few there, to raise public fears, just as the U.S. anthrax attacks caused alarm and panic in many places, but killed very few. Mass murder brings terror, but random murder is more terrifying.

The Aum Shinrikyo sarin attack in the Tokyo subway in 1995 killed 12, but 1,000 more were affected. The 2003 invasion of Iraq, supposedly over chemical and biological weapons, saw none used by the Iraqis. It appears the world's armies are wary of using poison gas in open warfare, though it seems to be more acceptable to use gas against those unlikely to retaliate in kind. When Saddam Hussein's forces used gas against Kurds in northern Iraq, survivors who took refuge in Turkey were not examined for six weeks, but doctors found evidence of a vesicant like mustard gas and, from the survivors' descriptions, concluded that a nerve agent was also used. Later, traces of mustard gas were found at some of the villages.

Gases and nerve toxins can have devastating effects, but the days of genocide through poison are now gone, for the single reason that it would be impossible to cover up such an event. A far more insidious and secret poisoner may be bacteria.

In their 2000 book *Plague Wars*, Tom Mangold and Jeff Goldberg claim that, during the apartheid era, the South African government's agents may have killed as many as 200 of its opponents with food and drink laced with bacteria. Somebody knew what they were doing, because bacteria and small animals have been poisoning large animals and humans far longer than humans have been using poison.

10

ENVENOMED FANGS AND STINGS

'The idea of using a form of poison which could not possibly be discovered by any chemical test was just such a one as would occur to a clever and ruthless man who had had an Eastern training. The rapidity with which such a poison would take effect would also, from his point of view, be an advantage. It would be a sharp-eyed coroner, indeed, who could distinguish the two little dark punctures which would show where the poison fangs had done their work.'

Sherlock Holmes, in Arthur Conan Doyle, *The Speckled Band*

Primo Levi was taken with the way an insect "in a brain weighing the fraction of a milligram . . . can store the crafts of the weaver, the ceramicist, the miner, the murderer by poison, the trapper and the wet nurse." Harm is done automatically, and the venomous animals have no need to think, because venom is generated at the molecular level.

Almost every group of animals, other than the birds, has at least one member that carries a venom. The male platypus is the only mammal, and the Gila monster may be the only lizard, but the rest of the animal kingdom is scattered with individuals with a poisonous defense. The Crown of Thorns starfish, cone shells, the cane toad, and a Brazilian caterpillar, *Lonomia obliqua*, deadlier than most snakes, are all out there, along with many others.

The *Oxford English Dictionary* lists three interesting words— well, it lists many, but three that are germane to this book: *toxicomania*, a morbid craving for poisons, and *toxiphobia* or *toxico-phobia*, both meaning "a fear of poisons." As a rule, toxicomanes are rare, if only because they tend to disappear soon after discovery, but toxicophobes last rather longer.

Australia, regarded by many of the fearful as the home of fearsome animals, sees 1,800 road deaths a year, 120 knife and 60 gun deaths, four from snakes, two each from bees, sharks, and lightning, one from crocodile attack, and none from spider bites. Of course, before antivenoms were developed, there were many more poison deaths, but snakes have traditionally fascinated. In the nineteenth century, the rattlesnake captured the morbid imagination of Americans, but there are now about 12 snake deaths each year in the United States.

Exotic Indian snakes such as the one Holmes refers to in *The Speckled Band* sent frissons of fear down the spines of nineteenth-century Britons. Their fear was helped along by the writings of men like Sir Joseph Fayrer (1824–1907), who made a long and patient study of poisonous animals in India. A snippet caught my eye, a report in *The Lancet*, December 17, 1870, of 11,416 cases of snakebite, including "6,645 in Bengal, Assam and Orissa, 1,995 in the NW Provinces, 755 in the Punjab, and 1,205 in Oude." "Dr. Fayrer thinks there must be 20,000 deaths annually in the whole of Hindoostan."

Fayrer began as a naval surgeon, studying with T. H. Huxley, and probably met Edward Pritchard at this time, as both were attached at least nominally to HMS *Victory*, but by 1850, Fayrer was serving in Bengal. By 1859, he was professor of surgery at the Medical College of Calcutta, and he accompanied the then

Prince of Wales on a tour of India. His *Thanatophidia of India* (1872, 1874) was cited by Charles Darwin when he wrote on venoms and their effects on the carnivorous sundew plant, *Drosera*. Sadly, according to Holmes scholars, Conan Doyle seems not to have consulted Fayrer's work, so the snake in *The Speckled Band* does not appear to be a recognizable species.

It was Fayrer who persuaded Huxley to join the Royal Navy, setting him on the path that would bring him to Australia with John MacGillivray (appropriately, they sailed on HMS *Rattlesnake*), where he met the future Mrs. Huxley, with whom he would found a famous dynasty, while gaining fame as Darwin's bulldog and countering the venom of the lowbrow divines who were appalled by the awful ideas of evolution that Darwin was putting about. In 1866, Fayrer invited Huxley to India to explore linguistics with him, but Huxley declined and Fayrer turned to poisonous snakes in a work "illustrated in colour by Hindu artists," and made generations of British even more fearful of snakes.

Readers of Kipling's *Rikki Tikki Tavi*, or of Herford's poem "The Mongoos," would know at least one way to be secure from snakes: keep a mongoose. Of course, any number of islands have suffered ecological damage from the mongoose, but it is certainly effective at killing snakes when there are no endangered bird species to chomp, and this raises the question: how does the mongoose manage against the snake?

Snake venom varies, but as a rule it contains one of those poisons that interferes with acetylcholine, like Botox or nerve gases. Snakes have a differently shaped acetylcholine receptor, and so, it turns out, does the mongoose.

Snakes' venom does not protect them from all predators, however, and monitor lizards are partial to the occasional meal of snake. A monitor's presence was usually a good early warning sign of the presence of a more unwelcome reptile. Horatio, Lord Nelson, had one such encounter on his travels.

> He had ordered his hammock to be slung under some trees, being excessively fatigued, and was sleeping, when a monitory lizard passed across his face. The Indians happily observed the reptile; and knowing what it indicated, awoke him. He started up, and found one of the deadliest serpents of the country coiled up at his feet. He suffered from poison of another kind; for drinking at a spring in which some boughs of the manchineel had been thrown, the effects were so severe as, in the opinion of some of his friends, to inflict a lasting injury upon his constitution.

> Robert Southey, *The Life of Horatio Lord Nelson*, 1813

The manchineel, or mancanilla (*Hippomane mancinella*), claimed many unsuspecting victims. Basil Ringrose was a buccaneer in the Caribbean, and he wrote in his journal of an incident on the island of Cayboa in the Gulf of Panama in 1679. "I was washing myself, and standing under a mancanilla tree, a small shower of rain happened to fall on the tree and from thence dropped upon my skin. These drops caused me to break out all over my body into red spots, of which I was not well for the space of a week after."

The mancanilla is a tree reaching some 40 to 50 feet, mostly found on sandy seashores in South America, Venezuela, Panama, and the islands of the West Indies. Merely sleeping in its shade was said to be enough to kill, perhaps in the way Ringrose described. Five years later, John Esquemeling mentions the death of another and less lucky buccaneer, William Stephens: "It was commonly believed that he poisoned himself with mancanilla in Golfo Dulce, for he had never been in health since that time."

It was not just the strange trees and animals that travelers had to learn about the hard way, many times over. There was always the risk that food known to be innocuous in one place may not be so in another. Xenophon describes an unfortunate experience he and his troops shared when they tried honey that the industrious local bees had made from *Azalea pontica:*

Here, generally speaking, there was nothing to excite their wonderment, but the numbers of bee-hives were indeed astonishing, and so were certain properties of the honey. The effect upon the soldiers who tasted the combs was, that they all went for the nonce quite off their heads, and suffered from vomiting and diarrhoea, with a total inability to stand steady on their legs. A small dose produced a condition not unlike violent drunkenness, a large one an attack very like a fit of madness, and some dropped down, apparently at death's door. So they lay, hundreds of them, as if there had been a great defeat, a prey to the cruellest despondency. But the next day, none had died; and almost at the same hour of the day at which they had eaten they recovered their senses, and on the third or fourth day got on their legs again like convalescents after a severe course of medical treatment.

Xenophon, *Anabasis, c.* 360 BC

Forewarned is forearmed, but sometimes experienced advice was ignored. In this account, the *almiranta* is the second ship of a fleet, the *capitana* the flagship.

There was in the *almiranta* an honest sailor called Saabedra, very experienced in the coast of Havannah and New Spain who said to Luis Baes and to me 'Notice Gentlemen that much of this fish is jaundiced, namely that which has black teeth, and it is pure poison, do not eat it but throw it into the sea and only eat that which has white teeth;' this was done and having cleaned out some of those with black teeth they were eaten by two cats and two young pigs and they all died within two days. They gave this advice at once to the people of the *capitana*, but they called us gluttons, as if we wanted them for ourselves.

Pedro Fernandez de Quiros, *The Discovery of Australia*, 1607

De Quiros goes on to explain how the seamen on the *capitana* ate the poisonous fish, then sent a boat at midnight, calling for priest and surgeon,

. . . for all the men were prostrate on the upper deck, asking for confession because they were dying. They went and did their duty and the surgeon took a jar of oil and gave it to the sick to drink and they vomited the food. The remedy was opportune for if they had delayed it would have done the same as to the cats and pigs; this poison closes the ducts of the faeces and urine and at once produces dementia, and there were some who did not come to themselves for more than a fortnight.

Pedro Fernandez de Quiros, *The Discovery of Australia*, 1607

Poison from fish, and from seafood in general, is a major topic, and one where ideas are changing all the time. Take, for example, the puffer fish, with its poison known as tetrodotoxin (TTX). This is a potent neurotoxin. In technical terms, it blocks voltage-gated sodium channels on the surface of a cell membrane. The molecule mimics the hydrated sodium ion, enters the gate, and binds to it. A hydrated sodium ion lets go in nanoseconds, while TTX holds on for tens of seconds. It has the same organizational effect as jamming elephants in all the gateways of a football stadium.

The TTX effectively blocks all sodium movement, and a single milligram is enough to kill an adult human. Tetrodotoxin is exquisitely well-designed. The molecule fits very exactly, and binds to no less than six sites on the sodium ion channel. Treatment normally comes down to artificial respiration and gastric lavage with activated charcoal.

TTX is a highly efficient killer and strangely spread among living things. A number of fish—*Fugu*, *Tetraodon*, *Arothron*, *Chelonodon*, and *Takifugu*—all store TTX and related analog in their tissues, but TTX turns up also in the blue-ringed octopus, seastars, crabs, marine snails and molluscs, flatworms, ribbonworms, and even marine algae. On land, some frogs, newts, and salamanders carry the same venom.

The scattered distribution of such a perfect killer molecule in unrelated groups of living things says something to biologists. It goes against all probability that such a diverse group of animals (and even algae) could have evolved the same poison. What is far more likely is that some smaller organism, down at the bottom of the food chain, makes the toxin, which others collect; a theory supported by the fact that puffer fish grown in culture do not contain TTX until they are given tissue from a TTX-producing fish to eat.

A single-point mutation in the sodium ion channel is enough to make the puffer fish immune to TTX. The mutation clearly does not affect the operation of the channel too much, but it makes the fish immune to something that up until then had poisoned its environment. Current thinking identifies the culprits making the toxin as bacteria or something of a bacterial size range.

There are other small poisoners in the sea aside from the mysterious sources of TTX. Filter-feeding bivalves survive by straining tiny plankton from the sea as it washes past them. When there is an algal bloom, the bivalves can take on significant loads of whatever toxins the plankton happen to be carrying. These toxins are not a great problem for the bivalves, however, since they have eons of evolutionary filtering behind them. In fact, the toxins can be a definite advantage, because casual shellfish eaters such as humans do not have the same immunity.

Shellfish poisoning takes several forms: paralytic shellfish poisoning is self-explanatory, and, as you might expect, the toxins work by binding to sodium channels in nerves. Interestingly, the target sites are the same as for TTX. The diarrheic form is triggered by toxins from dinoflagellates, and it usually sets in within 30 minutes, causing cramps, chills, nausea, and diarrhea. It appears to be nonfatal, although the toxins may cause stomach tumors in the longer term. Then there is amnesic shellfish poisoning caused by domoic acid, a toxin from diatoms that can also be found in fish and crabs in some localities. In this case, the memory loss that gives this form of poisoning its name can last up to several years.

Dinoflagellate toxin can be found in many shapes, sizes, and species. Ciguatoxin, for example, is a secondary reason for avoiding moray eels whenever possible: apart from their bite, their flesh is loaded with this poison. Just a few years ago, nobody knew where the toxin came from. Some fishermen believed it developed in the intestines of fish that were not processed immediately, but fish that had been frozen immediately after they left the water were still found to be toxic. Another theory suggests the ciguatoxic fish became toxic by eating puffer fish, but ciguatoxin and tetrodotoxin are quite different.

It is now generally held that a single-celled organism, a dinoflagellate called *Gambierdiscus toxicus*, puts the toxin into circulation through the food chain. In 2001, after ten years of work by more than 100 researchers, the full chemical process of ciguatoxin synthesis, all 90 steps of it, was completed, a process that, in passing, was laid at the door of this tiny organism. Ciguatoxin claims 20,000 victims each year, and the toxin is known from more than 400 fish. The distribution of the toxin can be surprisingly patchy, with ciguatoxic fish sometimes coming from one side of a small island, but not the other.

A poisoned food chain lay behind a peculiar disease found only on the island of Guam, east of the Philippines, halfway between Japan and Australia. In the early 1950s, the local population was suddenly struck with a form of amyotrophic lateral sclerosis (ALS), or Lou Gehrig's disease, at a rate roughly 100 times the global average. The symptoms included paralysis, tremors, and rigidity, similar to parkinsonism, and an Alzheimer-like dementia, so the condition was dubbed parkinsonism-dementia complex (PDC) or ALS-PDC. There was no single standard form of the condition, and no two cases were the same.

The local Chamorro population called the disease lytico-bodig, *lytico* being paralysis, and *bodig* parkinsonism. In recent times, lytico-bodig has been dying out, threatening to disappear before a cause could be identified. Once more, an unexplained disease might disappear, only to lie low and break forth again, perhaps in a more dangerous form, somewhere else. More importantly, lytico-bodig might hold some of the clues that would allow medical workers to unravel the secrets of Parkinson's disease and other similar conditions.

A neurotoxin in the food supply was the main suspect, because the disease affected only Chamorros. Cycad seeds were the most likely source of the neurotoxin. In the traditional Chamorro diet, the seeds of *Cycas* sp.,[*] a tree native to Guam, are ground into flour called *fadang* or *federico*. The Chamorros are well aware of the toxicity of the seeds and rinse the flour several times before cooking it. The washing is an old practice, and Oliver Sacks found a reference to it in the works of Louis de Freycinet, a French explorer who called in at Guam in 1819. By all reports, *fadang* is delicious, but why was lytico-bodig only a problem among the Chamorro of Guam, and even then, only among the traditional people of Umatac?

---

[*] The use above of "*Cycas* sp." is deliberate: what was once *Cycas circinalis* is now known as *Cycas micronesica* in the *Cycas rumphii* complex, and all three names may be encountered in a single document.

Guam is part of Micronesia, a region of the Pacific where people from different islands are immediately recognizable. This is a consequence of most populations originating in small bands, together with later rises and falls in numbers. These genetic distillations, known technically as the "founder" and "bottleneck" effects, can produce populations where large numbers of members share rare genes. In this part of the world, therefore, it is quite possible lytico-bodig could be inherited in some similar genetic mischance, but there was still a nagging suspicion that the cycads were somehow involved. Straight genetics could be ruled out, as it seemed to be only the elderly who were developing the disease. This was another puzzle: as the old sufferers died out, they were not being replaced by younger victims—almost nobody born after 1960 developed lytico-bodig. Over time, mineral deficiencies and parasites were ruled out, leaving neurotoxins as the prime suspects, but working out how these poisons entered the diet took a little longer.

Hunting is part of the Micronesian culture, and a wander in the rainforest or the mangroves on any island is all too often accompanied by the crack of rifles. At the end of the Japanese occupation, after World War II, America took over and the area became known officially as the Caroline Islands. Rifles were readily available, and far more efficient than the snares the Chamorros had previously used to catch food. Prized delicacies such as fruit bats were now available at the squeeze of a trigger. Until, that is, the bats were almost entirely shot out. One of Guam's two bat species vanished by the mid-1970s; the other dwindled to fewer than 100 individuals, and Guam bats were taken off the menu.

People were still eating bats, but now they were imported from Samoa, and as the local bats grew scarcer, so did cases of lytico-bodig. Conclusion: there was something in the local bats where occasional eating caused no real harm but the more often they appeared on the menu, the more people got sick. Whatever it was, the bats from Samoa did not have it—and Samoa is cycad-free.

So now it appears the mystery disease was a simple case of poisoning, but that still leaves the problem of identifying the toxin that the Guam bats held in their flesh. It may well have come from the cycads, but the bats would have to be bioaccumulating or building up its levels in their bodies. The most likely suspect is a known neurotoxin, cycasin, but it is water-soluble and so unlikely to accumulate in the bats' tissues. Whatever it was, the unfortunate bats of Guam may have left a spiteful legacy to their predators. It was just the bats' bad luck that the legacy was not associated with a warning color or bitter taste that might have educated the predators sooner.

Roger Ascham called his 1545 treatise on archery *Toxophily*, meaning "the love of archery." The similarity between toxophily and toxic is not mere chance, and the overlap starts with the Greek word, *toxon*, meaning "bow," from which we got *toxicon*, "having to do with the bow," in the form of *toxicon pharmakon*, poison for putting on arrows.

Hunters throughout the ages have taken a leaf out of nature's book and used poison as a weapon, because their prey would remain maddeningly out of reach of conventional weapons. All too often a tasty meal would last be seen scampering up a trunk and out onto a slim branch, where it could snigger derisively at its pursuers. A pursuer who had a poison dart or a poison arrow, however, only needed to wing his target, then wait for it to fall off that safe branch and into range of a big stick.

The only drawback for the hunters was the risk that some of the poison might remain in the meat, but, with luck, the poison's effect was in some way reduced in passing through the animal or in cooking, or maybe the toxin was less poisonous when swallowed.

As well as worrying about the food their local guides provided for them, travelers and explorers always feared these locals using poison weapons against them. William Dampier sailed the northern coast of Australia in the late seventeenth century, and wrote in 1699: "My young man, who had been struck through the cheek by one of their lances, was afraid it had been poisoned, but I did not think that likely. His wound was very painful to him, being made with a blunt weapon; but he soon recovered of it." Sir Joseph Banks, some 70 years later, had much the same fears on Australia's east coast. Here he describes, in his usual eccentric spelling, how they came on an Aboriginal camp, and how it was looted:

> We however thought it no improper measure to take away with us all the lances which we could find about the houses, amounting in number to forty or fifty. They were of various lenghs, from 15 to 6 feet in lengh; both those which were thrown at us and all we found except one had 4 prongs headed with very sharp fish bones, which were besmeard with a greenish colour'd gum that at first gave me some suspicions of Poison. . . . Upon examining the lances we had taken from them we found that the very most of them had been usd in striking fish, at least we concluded so from sea weed which was found stuck in among the four prongs.
>
> Sir Joseph Banks, *Journal*, 1770

There was, however, one part of the world where deadly poisons were indeed used as a regular thing by hunters, and this was South America, though the story of how Europeans discovered this is a little convoluted. Certain French scientists, perhaps more out of bloody-mindedness than any exercise of logic, argued that the Earth was like a rugby football on its end, as opposed to the logical English view, put forward by Isaac Newton, of our world as an oblate spheroid, a bit like a sat-upon and under-inflated soccer ball.

Newton's model arose from careful analytical thought, but the French were determined to prove their theorem. [*HYPOTHESIS* handwritten] To do this, they needed to measure a degree of latitude as far north as possible, and again as close to the equator as possible, so two expeditions were sent forth. One went to Lapland, and had some creditable scientists among it who did creditable things, but the one that went to South America was a mixed bunch, and some of them did incredible things. One such was Charles-Marie de la Condamine, who would spend ten years in South America in all.

During this time, Condamine saw tribesmen using blowpipes to hunt, and inspected and analyzed the poisoned arrows, reporting that the poison was "so active that, when it is fresh, it will kill in less than a minute, any animal whose blood it entered." This was curare, and, in another time, its discovery might have changed warfare. As it was, Condamine made a more influential discovery for warriors, one that provided quiet boots and excellent non-slip knife handles for commandos, suits for frogmen and tires for motor transport. In short, he discovered *caoutchouc*, as he named it, but it was given a more practical English name by Joseph Priestley. Based on its use to rub pencil marks off paper, he called it rubber. Curare came back from South America to Europe at the same time. Eventually, it would be used in medicine, even as the Amazonian hunters continued using it for its original purpose. It was left out of the armory, however, because there were much better poisons to use in war.

Many of the indigenous hunters around the world used stupefiers, a practice that is now being used to devastating effect by modern-day hunters seeking live fish for the gourmet seafood trade. They squirt cyanide into a coral reef in order to stupefy fish and catch them alive, but the poison hits more than the fish. Slowly,

A "hunting and gathering" poison from the rainforest can now be found in the suburban garden as a pesticide. In Central and South America, rotenone, in the form of extracts of the ground-up root of Jamaica dogwood (*Piscidia erythrina*) and other plants, has been widely used to poison fish. Rotenone kills insects as well, and it has been used in agriculture to control lice, fleas, and assorted larvae. It seemed to be safe when swallowed by birds and mammals, and it has even shown some interesting anti-cancer effects.

Problems can arise when it is sprayed, however, because inhalation of rotenone is by no means the same thing as swallowing it. The digestive tract is much more ferocious to chemicals than the lining of the lungs, which may usher the offensive molecules gently into the bloodstream, unchanged. Recently, when rats were regularly injected with rotenone over a period of some weeks, they developed symptoms resembling those of Parkinson's disease, including difficulty in walking and shakiness of the extremities—and their brain cells showed some protein deposits. It remains to be seen whether this means rotenone is more dangerous than we thought, or whether it simply causes rats to mimic parkinsonism, but there is every chance this discovery may be a boon to medical researchers if it gives them a way of inducing the symptoms they seek to cure.

piece by piece, the reef is killed by this sort of "hunting." The original users of this method would only have used stupefiers to stun the fish in a stream, without any lasting harm to the ecosystem.

The use of poisons to catch things remains normal practice for entomologists, although they are not so concerned with their catch being alive. They will sometimes spread sheets under a rainforest tree, for example, and fog it with insecticide, so thousands of dead insects rain down, adding a little more to our knowledge of the many species that are being endangered as the rainforests disappear. I intend no irony here: the lost insects will soon be replaced by others of their species in other trees, and there is really no other way to know what species are there, other than by dislodging them.

The world's rainforests are dying in many ways: in some parts of the Amazon basin, the main cause is road-building for the benefit of hunters of gold, one of the few metals that seems to be entirely nonpoisonous. The yellow metal is almost inert, but of the two most common methods of extracting it from the environment, one uses cyanide while the second uses mercury. Both poison rivers in rainforests and elsewhere.

The gold miners who use mercury rely on the process of amalgamation, in its original meaning, where other solid metals are dissolved into liquid mercury. When the amalgam is later heated, the mercury evaporates, leaving behind the gold or other absorbed metal. In theory, all of the evaporated mercury is condensed and recovered—in reality, mercury escapes into the ecosystem at every stage in the process.

If gold seems to be nontoxic, the same cannot be said for silver, and there is even a name for the condition it causes. "Argyria" is a blue-black deposit of silver in the skin that can be localized in areas where repeated contact is made with silver dust or silver salts. Occasionally, silver can be inhaled, but however absorbed, it often ends up in the eye as a gray-blue layer of silver sulfide, or as the aforementioned skin discoloration. Either way, once it gets in, it does not get out again. These days, the most spectacular cases are seen in people taking "colloidal silver" as an alternative medicine. Other victims make silver nitrate or silver decorative objects, or engage in silver mining and plating.

As we saw in the Guam bats, animals can also acquire poisons, and monarch butterflies have a well-known trick of eating milkweed, *Asclepia syriaca*, which is full of cardiac glycosides. The monarchs can tolerate this, but the bluejays that might otherwise eat them find the butterflies repulsive, on account of the diet they chose as caterpillars.

Cardiac glycosides are found in oleanders and foxgloves, and they include the useful drug digitalis. These are classed as cardenolides, but there is another subclass known as bufadienolides. This class can be found in a number of South African plants, like the Cape honey flower (*Melianthus comosus*) and Cape tulips (*Homeria* spp.) and also the mother-of-millions (*Bryophyllum* spp.), introduced to Australia from southern and eastern Africa and Madagascar.

While poisons often teach us stern lessons, we can still be slow to learn. Even though *Bryophyllum* became a noxious weed in Australia and poisoned livestock over large areas, it is commerically available in the United States and Canada as an ornamental plant, despite being recognized as invasive. It can only be a matter of time before it causes problems. The plant kills grazing stock, pulling it out of the ground doesn't kill it, and burning it releases toxic fumes.

And then there is the scarlet-bodied wasp moth of Florida. These moths take nine hours to mate, and during this time, the entangled couple would make a tasty morsel for a bat or spider, which probably accounts for the male moth's ploy of dining healthily on the poisonous dogfennel plant before mating. Actually, he doesn't so much dine as vomit on the plant, dissolving the fibers into a nice toxic soup that he absorbs into a pouch on his front. Now he is safe from predators, but his mate is not.

Picture a Lepidopteran Lothario, tanked up with toxic juice and in search of a mate. Soon, he finds one, but before the serious business starts, he showers her with toxins that she, in turn, later passes on to the eggs, thus protecting the next generation as well. No spider or bat wants to know about such a toxin-spattered morsel.

Toxins can play another role in mating, one that shows the strange logic that sometimes applies in the evolutionary game, where anything goes, as long as it results in more offspring. Fruitfly semen contains chemicals, some similar to spider toxin, that induce egg production in females and delay subsequent mating with another male. The effect of this is to make it more likely that a given mating will produce fertilized eggs with the toxic male's genes.

There is just one minor snag: the chemicals also reduce the female fruitfly's life span. But as long as she has laid *his* eggs by the time she dies (and she will have), the male has no concerns about her shorter life. There has even been an elegant experiment, where females and males are bred monogamously for a few generations and, by simple selection, the semen ceases to be toxic. The male's genes get an advantage from toxic semen in a promiscuous situation, but in monogamy the male with non-toxic semen produces more offspring from his single female. Under those conditions, the cost of the toxic semen outweighs the advantage it confers.

You could be excused from thinking that very little in the natural world could be smaller or more lethal to its recipient than toxic fruitfly semen, but the world's best poisoners are very, very, very small. There is no place to hide from these poisoners.

11

THE TINY
POISONERS

Up the airy mountain
Down the rushy glen,
We daren't go a-hunting,
For fear of little men . . .

William Allingham, *The Fairies*, *c.* 1870

These days, of course, we still believe in invisible things that make us ill, but now we call them germs. There are other, almost invisible things that make us ill, invertebrates that feed on us and spread disease, and some that harm us with their stings, but germs do it without any equipment at all. The poisoning game is one even the tiniest can play on equal terms, so even if we are too clever to fear little men anymore, we should fear the tiniest organisms for their toxins.

We assume hygiene to be a recent invention, but consider this description of food poisoning, taken from the Prologue to the "Cook's Tale" in Chaucer's *Canterbury Tales*. It is given here once in the original, and once in a later version that largely retains the feel of Chaucer's language while making more sense for modern readers. The cook asks if the other pilgrims will give him a hearing:

Oure Hoost answerde and seide, 'I graunte it thee	Our Host answer'd and said; 'I grant it thee
Now telle on, Roger, looke that it be good;	Roger, tell on; and look that it be good,
For many a pastee hastow laten blood,	For many a pasty hast thou letten blood,
And many a Jakke of Dovere hastow soold,	And many a Jack of Dover hast thou sold,
That hath been twies hoot and twies coold	That had been twice hot and twice cold
Of many a pilgrym hastow Cristes curs,	Of many a pilgrim hast thou Christe's curse,
For of thy percely yet they fare the wors,	For of thy parsley yet they fare the worse,
That they han eten with thy stubbel goos;	That they have eaten in thy stubble goose:
For in thy shoppe is many a flye loos.'	For in thy shop doth many a fly go loose.'

Geoffrey Chaucer, *Canterbury Tales*, *c.* 1387

Two things stand out in the comments of the Host here: the risk of reheating cold food that may already be tainted, and the part flies play in spreading illness in some unspecified form. The cook was in the habit, he says, of serving reheated Jack of Dover, taken to be warmed-up pie, of "letting the blood" (draining the gravy) from pasties, his parsley makes them ill, and flies are loose in his shop. Duncan Gow's children made him a "parsley sandwich" that was in fact hemlock, but this is probably not Chaucer's meaning here.

The world of science and medicine has two opposing views of disease, the environmental and the microbial. Malaria, for example, was first thought to be caused by the miasma of swamps and later by tiny life forms, injected into the bloodstream by mosquitoes. The evidence for swamp miasmas was that draining the swamps reduced the incidence of malaria, but we now recognize that draining the swamps reduced the number of mosquitoes. Long before the emergence of the germ theory in the 1860s—at least as far back as the eighteenth century and maybe even longer—a few people had at least an inkling that there were tiny things causing some sorts of illness, maybe not Allingham's "little men" but something too small to be seen by the naked eye:

> [I cannot believe that people] . . . talk of infection being carried on by the air only, by carrying with it vast numbers of insects and invisible creatures, who enter into the body with the breath, or even at the pores with the air, and there generate or emit most acute poisons, or poisonous ovae or eggs, which mingle themselves with the blood, and so infect the body.
>
> Daniel Defoe, *Journal of the Plague Year*, 1722

Defoe is best known to us as the author of *Robinson Crusoe*. Here he was probably relying on notes kept by his uncle, Henry Foe, who actually lived through the London epidemic, since Defoe had to have a source for his vivid descriptions of life in the time of plague. So in the 1660s or, at the very latest, around 1720, somebody must have been gossiping in pub or coffee shop about microbes, sufficiently often for one of the Defoe family to pick up on it.

The people who might be tempted to try Botox are the same people who will spread their kitchens and bathrooms with hideous poisons, designed to kill every germ, because germs are seen as an even more deadly form of poison, while disinfectants are good, and harmless. The ideas of "poison" and "germ" are intertwined in history, and certainly in this tale, first published by Oliver Wendell Holmes in the 1840s:

And if this were compared with the effects of a very minute dose of morphia on the whole system, or the sudden and fatal impression of a single drop of prussic acid, or, with what comes still nearer, the poisonous influence of an atmosphere impregnated with invisible malaria, we should find in each of these examples an evidence of the degree to which nature, in some few instances, concentrates powerful qualities in minute or subtile forms of matter.

Oliver Wendell Holmes, *Medical Essays*, 1842

In the same vein, Charles Darwin (who may or may not have treated himself with arsenic when he was young) wrote in *The Voyage of the Beagle* that "at this period the air appears to become quite poisonous; both natives and foreigners often being affected with violent fevers." On the question of Darwin's treatment, some people claim that there is no proof of it, but he was certainly treated with calomel, and there was this letter written in 1831:

Ask my father if he thinks there would be any objection to my taking arsenic for a little time, as my hands are not quite well, and I have always observed that if I once get them well, and change my manner of living about the same time, they will generally remain well. What is the dose?

Charles Darwin, letter to Miss Susan Darwin, September 6, 1831

When we look at the large number of bacteria that make us ill or kill us by the production of assorted toxins—ranging from diphtheria to whooping cough to food poisoning, to cholera, tetanus, and more—it is quite legitimate to think and speak of the bacteria as poisons, at least indirectly. In a few cases, those toxins they generate are quite capable of causing the symptoms of the disease even in the absence of the bacteria, so it is no wonder that people were willing to hang on to the poison model a little longer, if they were so inclined. Old ideas, old paradigms always take a while to die out completely.

In this case, the decisive period was, roughly, the decade of the 1860s. The medical world was beginning to take the germ theory seriously, and to look ever more closely at the chemical poisons that might be used against those other poisons, now revealed as microbes, to make people healthy again. The process of change was still incomplete in 1875, when Alfred Taylor made it clear that, for him, rabies was still a poison, not a microbe: "arsenic is a poison whether it enters the blood through the lungs, the skin, or the stomach and bowels: but such poisons as those of the viper, of rabies, and of glanders, appear to greatly affect the body only through a wound in the skin." (The horse—and, occasionally, human—disease glanders is caused by a bacterium called *Pseudomonas*, but in 1875 this bacterium was yet to be seen.) In time, new culturing methods, oil immersion lenses using toxic oil of savin, and new artificial dyes that stained bacteria by binding tenaciously to them would change all that. Later still, these dyes would be used to attach poison molecules to the bacteria, but that is another story.

Perhaps the best recorded example of the way thinking developed over the decade can be seen in the reports of the Royal Commission that investigated the outbreak of "cattle plague" that began in the summer of 1865—these days, we would call it rinderpest. It seemed a novel event, because the last outbreak of rinderpest in Britain had been over a century earlier.

The 1865 plague showed up first at Islington cattle markets in June, but within a month it had reached East Anglia, Shropshire, and Scotland. Transport has always been a major factor in the spread of epidemics: road transport in Africa carried HIV away from its starting point; the bubonic plague outbreak on the Pacific rim around 1900 owed its start to steam trains in China and steamships to places like the east coast of Australia and the west coast of America; and, in more recent times, we have seen SARS and West Nile disease spreading with help from jet aircraft. In 1865, a combination of free trade and railway cattle trucks allowed the cattle plague to spread far and fast, especially as nobody knew how it was spreading.

By September, 1,702 farms and 13,000 cattle were affected. By January 1866, 120,000 cattle were affected, and the situation was getting worse. The authorities imposed a weak ban on moving cattle, but there was no effective enforcement of the ban, so cattle were still moved and the disease still spread. People might have been gaining a clearer understanding of infection, but its causes were still a mystery. The London *Times*, whose editorials were more influential than any broadsheet or its editor today could dream of, was opposed to the notion of contagion, which meant people tended to favor miasmatic theories of plague poisons spreading far and wide.

That would all change, of course, when Robert Lowe, a member of the Royal Commission, became an editorial writer at the *Times* in October 1865. The tide of opinion began to change, but by then experience too had imposed a greater acceptance of the contagious nature of the disease, and, in February 1866, when the commission's second report came out, there was general agreement that only slaughter could stamp out the plague. In the end, half a million head of cattle died.

As science confronted the plague, there was a great deal to consider, and all the best minds were set to the problem—Lyon Playfair, William Crookes, J. B. Sanderson (after whom J. B. S. Haldane was later named), and a pupil of Justus Liebig named Angus Smith.

Despite the germ theory starting to gain some prominence, time and again we find references in the commission's report to the poison causing the disease: telling phrases such as "the blood contains the poison of the disease, so that serum obtained from it will give the disease by inoculation," and, "the poison contained in a minute portion of the mucous discharge . . . multiplies when it is injected into an animal, and so causes it to sicken in turn." On the other hand, they also spoke of disinfection, albeit in a qualified way:

> Disinfection, in the sense in which the word is used here, implies the destruction of an animal poison in whatever way it is accomplished. To find a perfect disinfectant for the Cattle Plague poison would be to stop the disease at once.
>
> *Third Report of the Commissioners into the Cattle Plague*, May 1866

In 1854, Angus Smith and one Alexander McDougall had patented McDougall's Powder for use as a sewage deodorant. McDougall manufactured and sold the powder, which was mainly carbolic acid, which Taylor describes as "a crystalline product of the distillation of coal-tar. When pure, it melts at 102°F. It has a characteristic, and not unpleasant odour." He adds that this poison could be as swift as prussic acid in its action, bringing death in as little as 20 minutes, and usually within four hours. Mostly used for suicide, he notes, it leaves white staining around the mouth and brown stains called eschars on the skin where the poison trickles from the mouth. In 1864 McDougall used it successfully to kill parasites of cattle on a sewage farm. When Joseph Lister heard of this, he came up with the idea of using the spray in antiseptic surgery, becoming the first of the modern Great Poisoners of medicine.

By 1867, an exultant Lister was able to write:

Since the antiseptic treatment has been brought into full operation, and wounds and abscesses no longer poison the atmosphere with putrid exhalations, my wards, though in other respects under precisely the same circumstances as before, have completely changed their character.

Joseph Lister, *British Medical Journal*, 1867

In January 1866, William Crookes suggested to Smith that they patent the use of carbolic and cresylic acid as a disinfectant. They probably would have been unsuccessful (in part because the mixture had already been described in Crookes' own journal, *Chemical News*). In any event, Smith declined. He was, after all, one of the commissioners on the Cattle Plague Commission, and so might have been accused of having a conflict of interest. They agreed, though, on the efficacy of this approach, and they urged others to use disinfectants.

Crookes and Smith seem to have thought that the disinfectants worked by dissolving the poison rather than killing the microbes, but, whatever their thinking, Crookes knew that he was on the right trail at last. As early as December 1865, he had advised a friend, a Mrs. Carmichael of Thirsk, to use carbolic acid as a disinfectant. Not content with this, he went to her farm to make sure the procedures he had recommended were carried out. All Mrs. Carmichael's 25 cattle were saved, but surrounding farms all suffered big losses. Smith and Crookes set out their method:

Wash the woodwork of the (cow)sheds everywhere with boiling water, containing in each gallon a wineglassful of carbolic acid. Then lime-wash the walls and roofs of the shed with good, freshly-burnt lime, adding to each pailful of whitewash one pint of carbolic acid. Cleanse the floors thoroughly with hot water, and then sprinkle freely with undiluted carbolic acid. Lastly, close all the doors and openings, and burn sulphur in the shed, taking care that neither men nor animals remain in the shed while the burning is going on.

Angus Smith & William Crookes, *Recommendations for Disinfection*, 1866

Crookes's ownership of *Chemical News* came in handy, as he was able to write up his procedures and results—and accord himself just a bit of self-praise—free of any unpleasant peer review requirements. He wrote a report that he serialized in the journal, saying the sheds were easily contaminated by a virus (a word that he used in the older sense of *poison*) like smallpox. He also carried out an experiment, of sorts, where he collected air from sheds housing dying cattle and passed it through cotton wool. He treated one half of the cotton wool with carbolic acid vapors for half an hour, and then inserted the two pieces in two calves. The calf that received the untreated cotton wool died, the other lived—hardly a defining result, but it was at least indicative. The official report of the Royal Commission carried the most weight, however:

A large number of substances which can be used in many other cases as disinfectants must be put aside. . . . Compounds of iron, zinc, lead, manganese, arsenic, sodium, lime, or charcoal powder, and many other substances, want the volatile disinfecting power; iodine, bromine, nitrous acid and some other bodies are too dear, or are entirely volatile, or are injurious to the cattle.

On full consideration, it appears that the choice must lie between chlorine, ozone, sulphur, and the tar acids (carbolic and cresylic). Two of these bodies, viz., chlorine, in the shape of chloride of lime, and the tar acids, have the great advantage of being both liquid and aeriform; they can be at once added to discharges, and constantly diffused in the air.

Third Report of the Commissioners into the Cattle Plague, May 1866

The report adds that there is evidence that chlorine, ozone, sulfurous acid, and the tar acids "all actually do destroy the Cattle Plague poison," but, in the long term, the greatest benefits were seen when people began to think in terms of poisoning the microbes causing disease.

It would be a while, though, before living germs would be fully accepted as the main cause of disease. Indeed, even now, there are people who assert loudly that all disease is caused by toxins. They are generally followers of an excellent medical man, Sir (William) Arbuthnot Lane (1856–1943), who, in his earlier days, introduced the use of metal plates, rather than wires, to join fractures. He also pioneered a treatment for cleft palate, but in 1925 he went off the rails, at least from the viewpoint of mainstream medicine. All disease, he averred, was caused by toxins created in the bowel.

The result of this was that during the late 1920s, Lane's toxin theory of disease led to surgical removal of large parts of the intestine to deal with vague complaints such as headaches, backaches, and depression. Colonic irrigation—a throwback to old-fashioned purging—first became fashionable among the wealthy and hypochondriacal. If all those bacteria in the intestine were producing toxins, then, without a home, they would be stymied. Off with the bowel, Lane cried!

Well, who can say? Attitudes change, and while there is no evidence for it, perhaps there is something in what Lane said. Once it would have been heresy to suggest that bacteria caused ulcers, which were obviously caused by some form of "poison" secreted in the intestines—now we know *Helicobacter pylori* causes them. Ideas change as we get more knowledge. Just recently, a group of researchers in Melbourne and London suggested that a possible cause of what they dub "the diabetes epidemic" may be a toxin produced by *Streptomyces* bacteria consumed in foods, such as potatoes, taken from the soil—making potatoes even more risky than their glycemic index and solanine levels might suggest—and the infection may extend to other root crops as well:

> *Streptomyces* species are ubiquitously present in soil and some can infest tuberous vegetables such as potatoes and sugar beet. Hence dietary exposure to a *Streptomyces* toxin could possibly cause repetitive pancreatic islet-cell damage, and so be diabetogenic in humans genetically susceptible to autoimmune insulitis.
>
> Zimmet, Alberti & Shaw, *Nature*, 2001

At this stage, this remains just a plausible and interesting hypothesis, one of many arising from the radical advances in both medicine and biology in the twentieth century. Several generations of biologists and medical researchers have been inspired by Paul de Kruif's *The Microbe Hunters*, first published in 1927, but even as they read it, they were entering into an era where the true heroes of medical science would be the poison hunters.

> Tasting the dorsum of frogs in the field has proved to be a sensitive test for the presence of pharmacologically active compounds in skin secretions. The presence of such compounds confers bitter, burning, or otherwise unpleasant taste. However, the possibility that it may confer posthumous authorship on papers such as this is a strong negative factor in its use.
>
> Neuwirth, Daly, Myers & Tice, 1979

Not all searches after new poisons have been quite so risk-laden, not all applications of poison have been with malignant intent, and many of the most unpleasant poisons have been quite stealthy—and in many cases, microbial. We can live with that, though, largely because other parts of the biosphere will usually provide an antidote to those poisons.

Some medical workers now argue that one supposed deficiency disease, kwashiorkor, is, in fact, caused by a mycotoxin. Stomach ulcers are now blamed on a bacterium. Of course, the bacterial infection is treated with a poison, an antibiotic, but it remains to be seen how many other conditions are caused by toxins produced by tiny life forms.

Some forms of heart disease may be caused by viral infections of the cardiac muscle, and some cases of hardening of the arteries may be caused by plaque-forming bacteria, though there is probably no toxin involved in this case. Cancers may be caused by toxins released by viruses, but some cancers may be treated by sending in modified viruses to target and kill the cancerous cells.

Cells that go wrong and start multiplying out of control normally recognize that they are doing the wrong thing and destroy themselves. We call this apoptosis, a poetic name drawn from the Greek for "falling leaves," and recent research has shown just how vital this process is. It defends us from cancers, abnormalities that are normally snuffed out by special suicide bags that take out rogue cells. A cancer only gets away when the body's ability to poison a suspect cell is lost.

Just as a hive of bees or a nest of ants is genetically programmed to dispose of a single sick member for the benefit of the rest of the individuals with the same genes, so our cells will destroy the cells that go bad, assuming the affected cells do not destroy themselves first. And because other cells with the same genes do better as a result, evolution reinforces this behavior at every step along the way.

Certain types of cancer, about 60 percent of all malignant tumors, arise because a gene known as p53 has somehow been knocked out or disabled. When it is operating properly, p53 stops cells from slipping into uncontrolled cell growth, a common feature in cancer. The American researcher Frank McCormick designed an adenovirus, a common cold virus known as ONYX-015, to take advantage of this missing gene.

An ordinary adenovirus contains a gene called E1b that disables the p53 mechanism, allowing it to attack a healthy cell. An adenovirus without E1b could not invade normal, healthy cells because it would not be able to disable the healthy p53 gene. On the other hand, tumor cells that lack the p53 gene should be an easy target for the virus. In other words, a disabled adenovirus would be a magic bullet, a smart bomb, an agent that can tell friend from foe, and wipe out the foe. The virus would enter the tumor cells and, with no p53 gene to inhibit it, would replicate continuously, ultimately causing cell death.

This, at least, was the theory behind McCormick's work, but laboratory studies showed that ONYX-015 replicates in tumor cells in which p53 remains intact. So if that part can go wrong, what else might the virus be doing? Is it a cause for worry? Apparently not. McCormick took into account some work reported in *Nature* in 1998 to offer a possible explanation of what is happening. The explanation involves a second gene, p14ARF, which appears to be mutated and defective in some tumor cells—it seems the damage to p14ARF may indirectly disable the p53 function.

McCormick reported that when normal cells are infected with ONYX-015, the p53 protein is produced, and the virus is shut down. In tumor cells that are missing p14ARF, despite their p53 genes being intact, this p53 production does not occur, leaving the cells open to adenovirus attack. Their strength, the missing gene that stops the cancerous cells from poisoning themselves, can be their downfall. (More recently, this work encountered some setbacks and has now been discontinued, but the facts remain the same.)

There is a somewhat older medical tradition of magic bullets that began with German immunologist Paul Ehrlich, who entertained a novel idea. The nineteenth century had seen the science of microscopy advance by leaps and bounds, as better lenses and oil immersion lenses using juniper oil combined with new and marvelous stains that locked on to specific chemicals found in some cells but not others. Now thin sections of tissue could be examined, and different cells could be distinguished. Ehrlich's notion was to find a stain that locked on to microbes but not to any cell in the human body, a stain that might attach to a lethal dose of poison or just to a single cell or cell type.

His 606th trial was on arsphenamine, better known today as salvarsan, $C_{12}H_{12}N_2O_2As.2HCl.2H_2O$ or dioxy-diamino-arsenobenzol-dihydrochloride. This became the famous magic bullet for syphilis. It was less than perfect, and it had a reputation for killing the patient, but then syphilis itself resulted in a horrid death from madness, often preceded by blindness. Mercury, in the form of intramuscular injections of mercuric succinamide, was a common treatment (and gave rise to J. Earle Moore's aphorism, "Two minutes with Venus, two years with mercury").

These days, however, salvarsan has been superseded by more benign poisons, the antibiotics prescribed for all sorts of other conditions and which are equally effective on the spirochetes of syphilis. No doubt, given time, though, some of these conditions will start to develop resistance to the common antibiotics, and we will be back where we started.

The magic poison bullet model still works, though, and one of the more effective modern variations has been very small doses of ricin, or at least the A chain of the toxin, the one that does damage when it gets inside a cell. Of course, without the B chain that opens the door to let it in, the A chain can do nothing, but when attached to an antibody, ricin A chains can be released into cultures of human cells infected with diseases such as HIV-1 in latent form. Other similar structures have been used on a number of cancers, and researchers are now working on modified forms of the A chain that should be even safer.

So are there still new poisons out there to be discovered, or only old poisons to be rediscovered? Most probably there are both. One of nature's Great Poisoners is ergot of rye, a resting stage in the life cycle of the fungus *Claviceps purpurea*. The ergot is more formally called the sclerotium, a formation that hibernates in cold climates, ready to start a new generation the following year. Ergot replaces the seeds of rye, producing a purple lump that looks to the French like a cockspur, or *ergot*. The ergot looks quite unlike the true grain, but it was so common people thought it was part of the rye plant, until the 1850s when the true nature of the ergot was understood.

Ergot is not something we think about each day, yet it is not all that rare and unusual. I could go into my street and find it two houses away. There is nothing to fear from it, even though it is deadly, or potentially so. There are about 35 species of *Claviceps*, most occurring on grasses, like my neighborhood species. All of them form a sclerotium and the same types of mycotoxins, as the poisonous alkaloids of fungi are usually called, but most of them never get near the human food chain.

Ergotized rye

These fungi produce four groups of alkaloids, including lysergic acid and lysergic acid amides (think LSD) and two related compounds, ergotamine and ergocristine. There is no LSD as such in ergot, but any one of the mycotoxins can trigger hallucinations as part of what is termed "convulsive ergotism." There is another form, "gangrenous ergotism," where the blood supply to fingers, toes, and limbs is cut off, leading to death of the flesh and loss of the affected part, or to infection and gangrene that can prove fatal.

Mycotoxins are high on the list of accidental poisons. A few of the mycotoxins, penicillin among them, are helpful antibiotics, diabolically toxic to bacteria and just mildly unpleasant to us, but some are quite deadly to humans, though it may not always be immediately apparent when a victim is being poisoned.

For example, liver cancer is an important public health problem in developing countries, where it may be ten times as common as in developed nations. As many as 76 percent of recorded cases of liver cancer occur in Asia. This might possibly be caused by genetics or by lifestyle, or by something in the environment. Exposure to the hepatitis B or C viruses, or to fungal aflatoxins, or both, seems to be the prime suspect. One obvious question is why the liver is such a common target. The answer is that the majority of the liver's cells, called hepatocytes, are constantly taking on foreign molecules, and often, losing the battle.

Aflatoxins are metabolic products of some species of *Aspergillus* (mainly *A. flavus* and *A. parasiticus*) and are among the most potent liver cancer compounds known. So who is most at risk? It depends where you live: the legal limit for aflatoxin in food is 4 parts per billion (ppb) in France and the the Netherlands, 15–20 ppb in Canada and the United States, and 30 ppb in India. A lot depends on how rigorously these limits are policed, how much of the food eaten is close to the limit, or on exposure to hepatitis B, because that virus can increase the risk of liver cancer arising from food contaminated with aflatoxin thirty-fold.

While health authorities look at ways of reducing aflatoxin exposure, the evidence would seem to suggest that it would do much more good to vaccinate the developing world's population against hepatitis B. These toxins can be found in the developed world in peanut butter made from untreated "organic" peanuts. Such contaminated crops are hard to sell in the developed world, however, so they are usually sold in the developing country of origin, or find their way into famine relief.

One ergot product, ergonovine, may produce abortions, either by mischance or deliberately, and some of its other toxins cross over into breast milk, affecting infants who are particularly vulnerable. University of Maryland historian Mary Matossian, among others, has made a strong case for ergot as a major shaper of European history, an effect that she argues is largely hidden by other events.

The problem with interpreting old records is that not only does ergot produce over twenty different alkaloids, but different strains of ergot produce different amounts of individual mycotoxins, so Russian ergot was less likely to affect women's fertility, even though the Russian strains had a higher overall level of alkaloids. And if the wind-blown *Claviceps* can vary across a geographical region like Europe, it can most certainly vary over time.

Of course, depending on the strain and the way it is treated, ergot can also be useful. In German, it is called *mütterkorn* or "mother's grain," recalling the way German midwives used it in the sixteenth century to help women in labor. The dosage had to be measured extremely carefully. Just the right amount of the purple grain would hasten contractions; a little more and ergot was an efficient abortifacient; a very little more and the woman suffered gangrene and convulsions.

So, given that ergot is variable in the poisons it produces, a number of scientists have looked at patterns of disease, and wondered if some of the things we interpret as arising from a modern "disease" might not be a result of ergotism or some other mycotoxin, acting either directly on the victims and killing them, or indirectly by suppressing their immune system, so that death was caused by otherwise serious but not deadly diseases.

The evidence for this has to come from patterns of mortality—looking at who died, where, and when. Mary Matossian argues, for example, that British "ague," commonly understood to be malaria, broke out at the wrong time of year for mosquitoes, and it is more likely to have been caused by toxins from ergot-infected rye. This is a reasonable hypothesis, but the disease might also have been caused by some other mycotoxin growing on fruit varieties that have since been discarded or a strain of fungus that no longer exists.

Other grains have their own fungi; wheat, for example, has *Fusarium*, but its toxins do not pass through mother's milk like the ergot alkaloids. It is possible that the high incidence of infant mortality throughout the ages could have something to do with ergot, either as direct poisoning from breast milk or as a result of immature immune systems being suppressed and rendered vulnerable. These days, the highest death rates from infectious disease occur in dry summers, but before about 1750, wet summers were the deadliest. One way to explain this is to blame the deaths on ergotism or other fungal toxins, but whatever the reason, there seems to have been a Great Poisoner about—it is even possible there never was such a disease as the Black Death.

Most epidemicists assume that the Black Death was bubonic plague, but in 2002 anthropologist James Wood and his colleagues argued that the data do not fit. They claim that one reliable record, English priests' monthly mortality rates during the epidemic, shows a forty-five-fold greater risk of death than during normal times, a level of mortality far higher than the rate usually associated with bubonic plague. We might see those death rates when a previously uninfected population first encounters a new disease, but the figures appear too extreme for the disease to have been bubonic plague.

Modern bubonic plague typically needs to reach a high frequency in the rat population before it spills over into the human community via the flea vector. Historically, epidemics of bubonic plague are associated with enormous die-offs of rats, but there are no reports of dead rats in the streets in the fourteenth-century epidemic. The pattern is not that of more recent epidemics when the cause has been confirmed as bubonic plague.

It was the single symptom of lymphatic swelling that led nineteenth-century bacteriologists to identify the fourteenth-century epidemic as bubonic plague, but the symptoms of the Black Death also included high fevers, fetid breath, coughing, vomiting of blood, and foul body odor, as well as red bruising or hemorrhaging of skin and swollen lymph nodes. Many of these symptoms may indicate bubonic plague, but they can appear in many other diseases, so diagnosis based on just one symptom is, at best, unwise.

Clue patterns can be seen by looking at *where* a disease strikes. Cold, dry areas, unsuited to ergot—places such as Iceland, northern Norway and Sweden, Finland, and large areas of Russia and the Balkans—escaped the plague entirely. Wood believes that, rather than being spread by animal and insect vectors, the Black Death was transmitted through person-to-person contact, like measles and smallpox. The geographic pattern of the disease seems to bear this out, since it spread rapidly along roadways and navigable rivers and was not slowed down by the kinds of geographical barrier that would restrict the movement of rodents.

According to Wood, we can only trace modern bubonic plague reliably back to the late eighteenth or early nineteenth century. Who knows when it first emerged, or how it appeared? While the bacillus may have mutated over time, it may also have had some outside assistance. Overall, the original "Black Death" seems to have carried off a third of the population, but the mortality pattern was very patchy, so perhaps there was some sort of immunosuppression caused by mycotoxins in the most common of foods—rye, or black, bread.

Significant support for the involvement of a mycotoxin in plagues can be seen if we correlate rainfall patterns and outbreaks of plague, where dry weather halts the plague. In the four or five years before the great plague outbreak in London of 1665–66, Samuel Pepys makes constant reference to plague incidents in London: it was around, but not severe. Then 1665 and 1666 were very wet—and plague erupted.

So was ergotism a problem at the time of the Black Death? The winter of 1340 was very severe in Leicestershire, with heavy snowfalls and rain. The following summer saw people suffering fits, pains, and an overwhelming desire to bark like a dog. In passing, this was presumably the origin of *barking mad*, though it begs the question: why did people not feel an urge to meow like cats, sing like linnets (or swans), or chirrup like sparrows? Again, in 1355, there was an epidemic of "madness" in England, with people hiding in thickets and woods from demons. And with the fear of demons came fears of witchcraft.

Witches and witch trials seem to have a lot to do with ergotism. The witch trials of Salem seem to have been ergot-involved, and trials in Scotland in the sixteenth century were centered on rye-growing areas, suggesting an ergot connection. Throughout the 1560s, as Europe's climate grew colder and damper, witch trials increased in number. In England at this time, the Essex marshland was being reclaimed and the new ground was commonly sown with rye, the only grain that could thrive in the sour soil.

The real trap for "witches" was that they were people who knew their plants, who had wort-cunning, and so could cure some conditions. Unfortunately, their superstitious neighbors assumed those with the power to cure disease must also have the power to cause it, and this brought about witch burnings. Today, we are more advanced, so we allow the modern-day occupants of the witch role to insure themselves, and then we fasten onto them like leeches in courts of law. (Biologically, a parasite that allows the host to live to be preyed on another day is seen as more advanced than a pathogen that kills the host outright.)

Ergot may even have been responsible for the division of the old Holy Roman Empire into the two regions that would become France and Germany

We know "holy fire" was active in the Rhine valley in AD 857, reducing populations, checking fertility, and causing social disruption. By AD 900, the Norsemen were in France and pressuring the Holy Roman Empire to the extent that Charles III was forced to abdicate, and the empire split into eastern and western halves. The Norsemen had little use for rye, unlike the Franks.

Ergot may have been responsible for other world events. In July 1789, just a few days after the fall of the Bastille and just as the rye crop was harvested, *La grande peur*, "The Great Fear," started, running from July 20 to August 6. It took the form of rumors that brigands were on the way to seize the peasants' crops. No brigands transpired, but the fear was so strong that the National Constituent Assembly met at Versailles and voted to abolish the *Ancien Régime*, the social and economic order that kept the ruling classes on top of the heap.

Matossian has found references to the rye crop in July being "prodigiously affected by ergot," and there is clear evidence of illness caused by bad bread. While it would probably be excessive to argue that the French Revolution was caused by ergot, it would appear the direction of the revolution may have been partly ergot-driven.

If that seems too extreme a claim, consider this: in 1722, Tsar Peter the Great had assembled a huge army, ready to sweep into Turkey. The Turks were unusually ill-prepared for war, because a tulip-growing craze was diverting the whole of Istanbul. The Russians numbered 20,000 at Astrakhan, a force large enough to drive Turkish forces out of the Ukraine, which would then be free to join Russia, and perhaps even large enough to take Istanbul.

There was just one catch: rye hay and grain were brought in to feed the army and its horses, but in August, the ergot in the grain struck. Horses went down with the blind staggers and, soon after, men began to fall victim to Saint Anthony's fire. More than 20,000 soldiers and civilians died in and around Astrakhan that autumn, and Russia's best chance to push back the Turks and gain a warm-water port from which to open up foreign trade perished with them.

To most people, ergotism is a disease of the past. It might have caused dancing madness in medieval Germany or made people barking mad in England at the same time, but it held no terrors in the modern twentieth century. After all, ergot had been tamed. Indeed, one of its toxins, ergonovine, had been selected, purified, and added to the pharmacopoeia, with small doses being used to reduce bleeding in childbirth. Some of the others were being looked at, and one that would shape a generation, LSD, would soon be hailed as a powerful tool in psychiatry. It was quite clear to all that Saint Anthony's fire had been quenched forever.

Then, one day in 1951, everything changed. On August 12, a small Provençal town, Pont Saint Esprit, was struck by a strange disease that at first seemed to be appendicitis, but the symptoms were not quite right. Two patients seen by Dr. Jean Vieu both had low body temperatures and cold fingertips, were babbling, and had hallucinations. When another patient turned up with the same symptoms the next day, Vieu consulted with two colleagues. Between them, they had 20 patients with the same symptoms. Many of them were also exhibiting violent behavior, apparently because of the hallucinations.

Straitjackets were rushed to the town to restrain the victims of this sickness, and terror spread as people learned of a demented eleven-year-old boy trying to strangle his own mother. People began to whisper of mass poisoning by the authorities. It was some time before people realized ergot lay behind the outbreak, because it was 130 years since the last case, and subsequent generations of farmers knew how to treat ergot: first, the rye seed is immersed in a 30 percent potassium chloride solution, so the ergots float away leaving the seed behind; then the field is deep-ploughed to bury any spores, and a different crop is planted the next year, breaking the cycle.

The one thing you don't do is sell the seed, untreated, for human consumption. Sadly, a farmer, a miller, and a baker had conspired to get rid of diseased rye by hiding it in wheat bread. As a result, 200 were made ill, 32 became insane, and four died, in part because of greed and perhaps in part because nobody really knew how serious the effects were.

It is worth speculating that global warming will change crop rotations and growth patterns, and will also change the ability of toxic fungi to flourish where they are now unknown. The effects of this change may be quite unexpected, but, as we have seen, we are prepared to poison our world, so long as there is a profit to be made. At this stage we cannot say exactly how global warming will affect us in the future, but affect us it will, and we can reliably expect the unexpected.

We would seem to be running the microbes a close second as the greatest of the Great Poisoners, and while they do little to make life easier for us, we seem to be doing plenty to make life easier for them.

Are we coming to the end of the poisons era, entering a time when we use genetics instead of poisons to cure our ills? I suspect that we may move to some new, hitherto undreamt of, classes of poisons, but no, we won't really move away from poisons as such: not until our microbial enemies stop using poisons, and that has been going on for a *very* long time.

EPILOGUE

The hand that signed the paper felled a city;
Five sovereign fingers taxed the breath,
Doubled the globe of dead and halved a country;
These five kings did a king to death.

Dylan Thomas, "The hand that signed the paper," 1936

Of course, you would never use poison, would you? Affirm it to me: put your hand up if you wouldn't use poison, wiggle your fingers for emphasis . . .

Liar. The very fact that you have fingers to wiggle betrays you. I claimed at the start of this book that we are creatures of poison, but poison is in our recent past, as well as in our distant past, and our fingers prove it, the very fingers that make us human. They were shaped by poison, a very special poison, applied cell by cell to a slab of flesh, a mere flipper, separating the slab into five strips that became the fingers of the hand, and the thumb. Apoptosis not only protects us from cancers, but it also shapes us, as in the formation of fingers.

The operations of apoptosis are exquisitely precise, and one of the 2002 Nobel Prizes went to Sydney Brenner, Robert Horvitz, and John Sulston for their discoveries concerning "genetic regulation of organ development and programmed cell death," which is the Nobel committee's preferred name for apoptosis. Most of this work orbited around the discovery that a worm, barely a millimeter long, is formed from 1,090 cells, but that *Caenorhabditis elegans* destroys 131 of those, to form an organism with exactly 959 cells.

The same principle applies to all living things: the careful use of poison, apoptosis, shapes us as we develop, and keeps us alive, yet only 30 years ago, we knew nothing of this. Poison, it seems, has not yet finished surprising us.

BIBLIOGRAPHY

I am a supporter of Project Gutenberg, and even when I have a print copy to hand, I prefer electronic searching for a vaguely recalled phrase. On top of that, these are easier for readers to track down, so I have cited Gutenberg versions where I know one exists. A number of the verse references point to websterworld.com, because, as the site's poetry editor, I put them there. All in all, I hope this list will help the curious reader to track down some of the oddities.

Alcott, Louisa. *Little Women*, 1867. Project Gutenberg lwmen13.txt, accessed 3/11/04.

Allingham, William. "The Fairies," *c.* 1870. *Webster's World of Poetry*, www.websterworld.com.

Austen, Jane. *Northanger Abbey*, 1818. Project Gutenberg nabby11.txt, accessed 3/11/04.

Austen, Jane. *Persuasion*, 1818. Project Gutenberg persu11.txt, accessed 3/11/04.

Banks, Sir Joseph (1744–1820). *Journal*, vol. II. Sydney: Trustees of the Public Library of NSW, 1962.

Barrie, James. *Peter Pan*, 1911. Project Gutenberg peter16.txt, accessed 3/11/04.

Bennett, George. "On the Toadfish (*Tetraodon hamiltoni*) of New South Wales," *NSW Medical Gazette*, March 1871, vol. 6, no. 1, 176–81.

Bentley, Ronald and Chasteen, Thomas G. "Arsenic curiosa and humanity," *The Chemical Educator*, 7(2), published on the Web, March 1, 2002, http://chemeducator.org/bibs/0007002/720051rb.htm, last accessed 3/11/04.

Blyth, A. W. *Poisons: Their Effects and Detection*. London: Charles Griffin and Co., 1884.

Brontë, Anne. *The Tenant of Wildfell Hall*, 1848. Project Gutenberg wldfl10.txt, accessed 3/11/04.

Burney I. *The Poison Hunter*. Wellcome News 1999. Issue 20, Q3, http://www.wellcome.ac.uk/en/1/hompropoi.html.

Cameron, Sir Charles. *Reminiscences of Sir Charles Cameron, CB*. Dublin: Hodges, Figgis & Co. Ltd., Publishers to the University, 1913, http://indigo.ie/~kfinlay/cameron/cameroncontents.htm.

Carboni, Raffaello. *The Eureka Stockade*. Project Gutenberg rkstk11.txt, accessed 3/11/04.

Carefoot, G. L., and Sprott, E. R. *Famine on the Wind*. Ontario: Rand McNally & Company, 1967.

Carroll, Lewis. *Alice's Adventures in Wonderland*, 1866. Project Gutenberg alice30.txt, accessed 3/11/04.

Castiglioni, Arturo. *A History of Medicine*, trans. E. B. Krumbhaar, 2nd ed. London: Routledge & Kegan Paul, 1947.

Chaucer, Geoffrey. Prologue to the "Cook's Tale" in *Canterbury Tales, c.* 1387. Project Gutenberg cbtls12.txt, accessed 3/11/04.

Chesterton, G. K. "Song Against Grocers," 1914. *Webster's World of Poetry*, www.websterworld.com.

Conner, W. E., Boada, R., Schroeder, F. C., González, A., Meinwald, J., and Eisner, T. "Chemical Defense: Bestowal of a Nuptial Alkaloidal Garment by a Male Moth upon its Mate," *Proc. Nat. Acad. Sci.* USA 97, 14406–11, 2001.

Crabbe, George. "The Parish Regist," 1807. *Webster's World of Poetry*, www.websterworld.com.

Cribb, A. B., and J. W. *Wild Food in Australia*. Sydney: Fontana, 1976.

Crossland, Robert. *Wainewright in Tasmania*. Melbourne: G. Cumberlege, Oxford University Press, 1954.

de Kruif, Paul. *The Microbe Hunters*. Rep., New York: Harvest Books, 2002.

Defoe, Daniel. *A Journal of the Plague Year*, 1722. Project Gutenberg, jplag10.txt accessed 3/11/04.

Dick, Oliver Lawson (ed.). *Aubrey's Brief Lives*, written by John Aubrey *c.* 1690. Harmondsworth: Penguin Books, 1976.

Dickens, Charles. *Hunted Down*, 1859. Project Gutenberg hntdn10.txt, accessed 3/11/04.

Dickens, Charles. *Pickwick Papers*, 1837. Project Gutenberg pwprs10.txt, accessed 3/11/04.

Doyle, Arthur Conan. *The Illustrated Sherlock Holmes Treasury*. New York: Avenel Books, 1976.

Emsley, John. "The Trouble with Thallium," *New Scientist*, August 10, 1978, pp. 393–94.

Emsley, John. *The 13th Element: The Sordid Tale of Murder, Fire, and Phosphorus*. Hoboken: John Wiley & Sons, 2000.

Gilbert, Michael. *The Oxford Book of Legal Anecdotes*. New York: Oxford University Press, 1989.

Goodman, Jonathan (ed.). *The Lady Killers: Famous Women Murderers*. New York: Citadel Press, 1991.

Graves, Robert, *Goodbye to All That*, 1929. Revised ed. London: The Folio Society, 1981.

Graves, Robert. "They Killed My Saintly Billy," 1957. London: Xanadu, 1989.

Gruber, Howard E. *Darwin on Man*. New York: E. P. Dutton & Co., 1974.

Hagger, Jennifer. *Australian Colonial Medicine*. Adelaide: Rigby Limited, 1979.

Haldane, J. B. S. "On Being One's Own Rabbit," in *Possible Worlds and Other Essays*. London: Chatto & Windus, 1927.

Hall, John. *Select Observations on English Bodies*, 1657, quoted by Joan Lane in *John Hall and His Patients*.

Hamill, Philip (ed.). *Murrell's Poisons*, 14th ed. London: H. K. Lewis, 1934.

Hamilton, Alice. *Hamilton and Hardy's Industrial Toxicology*, 4th ed./revised by Asher J. Finkel. Boston: J. Wright, 1983.

Hammer, Norman. *A Catechism of Air Raid Precautions*, 4th (revised) ed. London: John Bale Sons & Curnow Ltd., 1939.

Herodotus. *Histories*, newly translated and with an introduction by Aubrey de Selincourt. Harmondsworth: Penguin Books, 1954.

Holmes, Oliver Wendell. *Medical Essays*, 1842–1882. Project Gutenberg medic11.txt, accessed 3/11/04.

Home Office. *Medical Treatment of Gas Casualties (Air Raid Precautions Handbook No. 3)*. Canberra: repr. by L. F. Johnston, Commonwealth Government Printer, n.d., but based on the first edition of 1937.

Housman, A. E. "A Shropshire Lad, LXII," 1896. *Webster's World of Poetry*, www.websterworld.com.

Hudson, W. H. *The Book of a Naturalist*. London: Thomas Nelson & Sons, n.d., but before 1928, probably published in 1920.

Hurst, Evelyn. *The Poison Plants of New South Wales*, compiled under the direction of the Poison Plants Committee of New South Wales. Sydney: The Committee, 1942.

Jones, Ian. "Arsenic and the Bradford Poisonings of 1858," *The Pharmaceutical Journal*, vol. 265, no. 7128, pp. 938–39, http://www.pharmj.com/Editorial/20001223/articles/arsenic.html.

Jose, Arthur Wilberforce. "John Tawell," *Journal of the Royal Australian Historical Society*, vol. 18, pp. 31–43.

Keats, John. *The Complete Poems*, ed. John Barnard, 2nd ed. Harmondsworth: Penguin Books, 1978.

Koenig, Robert. "Wildlife deaths are a grim wake-up call in Eastern Europe," *Science* 287, March 10, 2000, 1737–38.

Lane, Joan. *John Hall and His Patients*, with medical commentary by Melvin

Earles. Stratford-upon-Avon: The Shakespeare Birthplace Trust, 1996.

Lapierre, Dominique and Moro, Javier. *Five Past Midnight in Bhopal*. New York: Warner Books, 2002.

Layton, Deborah. *Seductive Poison: A Jonestown Survivor's Story of the Life and Death in the People's Temple*. New York: Doubleday, 1998.

LeFebure, Victor. *The Riddle of the Rhine*, 1923. Project Gutenberg rrhin10.txt, accessed 3/11/04.

Leichhardt, Ludwig. *Journal of an Overland Expedition in Australia: From Moreton Bay to Port Essington, a Distance of Upwards of 3000 Miles, During the Years 1844–1845*, published 1846, available as Project Gutenberg file xpvld10.txt, accessed 3/11/04.

Levi, Primo. *Other People's Trades*. London: Abacus, 1991.

Loder, Natasha. "Chemists 'volunteered for nerve gas tests,'" *Nature* 404, March 30, 2000: 428–29.

Lucretius, Titus Lucretius Carus. On the Nature of the Universe, IV. Project Gutenberg natng10.txt, accessed 3/11/04.

Marks, Leo. *Between Silk and Cyanide—A Codemaker's War 1941–1945*. New York: Free Press, 1999.

Matossian, Mary Allerton Kilbourne. *Poisons of the Past: Molds, Epidemics, and History*. New Haven: Yale University Press, 1991.

Meharg, Andrew A., and Killham, Kenneth. "Environment: A Pre-industrial Source of Dioxins and Furans," *Nature* 421, January 21, 2003: 357–60.

Montaigne, Michel de. *Essays*, 1575, trans. Charles Cotton. Project Gutenberg mn20v11.txt, accessed 3/11/04.

More, Thomas. *Utopia*. Project Gutenberg utopi10.txt, accessed 3/11/04.

Motion, Andrew. *Wainewright the Poisoner: The Confessions of Thomas Griffiths Wainewright*. Chicago: University of Chicago Press, 2001.

Neuwirth, M., Daly, J. W., Myers, C. W., and Tice, L .W. "Morphology of the Granular Secretory Glands in Skin of Poison-dart Frogs (Dendrobatidae)," *Tissue and Cell* 11: 1979, 755–71.

Nickell, Joe, and Fischer, John F. *Crime Science*. Lexington: University Press of Kentucky, 1999.

Ottoboni, M. Alice. *The Dose Makes the Poison: A Plain-language Guide to Toxicology*, 2nd ed. New York: Van Nostrand Reinhold, 1991.

Pausanias. *Description of Greece*, trans. with commentary by J. G. Frazer. New York: Biblio & Tannen, 1965, vol. 1: 557–58, Bk.X, Ch. xxxvii.

Perutz, Max F. *I Wish I'd Made You Angry Earlier: Essays on Science, Scientists, and Humanity*. Cold Spring Harbor Laboratory Press: 10 Skyline Drive, Plainview, NY 11803–2500, 1998.

Plato. *The Last Days of Socrates*, New York: Penguin Books, 2003.

Pliny. *The Natural History of Pliny*, trans. Philemon Holland. New York: McGraw-Hill, 1964.

Plutarch. *Life of Flamininus*, trans. John Dryden. http://classics.mit.edu/ Plutarch/flaminin.1b.txt, accessed 3/11/04.

Prior, Matthew. "The Remedy Worse Than the Disease," *c.* 1715. *Webster's World of Poetry*, www.websterworld.com.

Riddle, J. M. *Contraception and Abortion from the Ancient World to the Renaissance*. Cambridge, Mass.: Harvard University Press, 1992.

Riddle, John M. *Eve's Herbs: A History of Contraception and Abortion in the West*. Cambridge, Mass.: Harvard University Press, 1997.

Sacks, Oliver. *The Island of the Colorblind*. New York: Knopf, 1997.

Sax, N. Irving. *Handbook of Dangerous Materials*. New York: Reinhold, 1951.

Selinger, Ben. *Why the Watermelon Won't Ripen in Your Armpit*. Sydney: Allen & Unwin, 2000.

Shakespeare, William. *Richard II*, Act III, scene ii, spoken by King Richard. Project Gutenberg 1ws1511.txt, accessed 3/11/04.

Shakespeare, William. *Romeo and Juliet*, Act I, scene iii. Project Gutenberg 1ws1611.txt, accessed 3/11/04.

Shakespeare, William, *The Merchant of Venice*, Act III, scene i. Project Gutenberg 1ws1811.txt, accessed 11/3/04.

Sherard, Robert Harborough. *The White Slaves of England*. London: James Bowden, 1897.

Singh, Jagvir, *et al.* "Diethylene Poisoning in Gurgaon, India," *Bulletin of the World Health Organization*, 2001, 79 (2), 1998, 88–95, http://www.who.int/docstore/bulletin/pdf/2001/issue2/vol.79.no.2.88-95.pdf.

Smiles, Samuel. *Industrial Biography*, 1863. 1st ed. Project Gutenberg inbio10.txt, accessed 3/11/04.

Smith, Grover (ed.). *Letters of Aldous Huxley*. London: Chatto & Windus, 1969.

Smollett, Tobias. *The Adventures of Peregrine Pickle*, 1752. Project Gutenberg thdvn10.txt, accessed 3/11/04.

Smollett, Tobias. *The Expedition Of Humphry Clinker*, 1771. Project Gutenberg txohc10.txt, accessed 3/11/04.

Smollett, Tobias. *Travels Through France and Italy*, 1766. Project Gutenberg ttfai10.txt, accessed 3/11/04.

Southey, Robert. *The Life of Horatio Lord Nelson*, 1813. Project Gutenberg hnlsn10.txt, accessed 3/11/04.

Stashower, Daniel. *Teller of Tales: The Life of Arthur Conan Doyle*. New York: Henry Holt & Company, 1999.

Sturgess, Ray. "Death at the hands of doctors," *The Pharmaceutical Journal*, 269: 2002, pp. 899–900, http://www.pharmj.com/Editorial/20021221/christmas/doctors.html.

Suetonius, Gaius Suetonius Tranquillus. *The Twelve Caesars*, trans. Robert Graves. Harmondsworth: Penguin Books, 1957.

Taylor, Alfred Swaine. *Taylor's Principles and Practice of Medical Jurisprudence*,

3rd ed., ed. Thomas Stevenson, MD. London: J. & A. Churchill, 1883.

Taylor, Sherwood. *The World of Science*. London: William Heinemann, 1950.

Tewksbury, Joshua J., and Nabhan, Gary P. "Seed Dispersal: Directed Deterrence by Capsaicin in Chillies." *Nature* 410, 2001, p. 415.

The Third Annual Report of the Acclimatisation Society of New South Wales, 1864, held at the Australian Museum, Sydney.

Third Report of the Commissioners Appointed to Inquire into the Origin and Nature, &c. of the Cattle Plague; with an appendix. Presented to both Houses of Parliament by Command of Her Majesty, May 1866, http://post.queensu.ca/~forsdyke/rindpst2.htm, accessed 3/13/04.

Thomas, Dylan. *Under Milk Wood*. London: Aldine/J. M. Dent, 1954.

Thomas, Dylan. *Miscellany*. London: Aldine/J. M. Dent, 1963.

Thompson, Benjamin, Count Rumford. "Of Food and Particularly of Feeding the Poor," in *Essays, Political, Economical and Philosophical*, 1796. Project Gutenberg essbr10.txt, accessed 3/11/04.

Thompson, C. J. S. *Poison Mysteries in History, Romance and Crime*. London: The Scientific Press, 1923.

Tilden, Sir William. *Chemical Discovery and Invention in the Twentieth Century*. London: George Routledge & Sons, n.d. (preface dated October 1916).

Timbrell, J. A. *Principles of Biochemical Toxicology*, 3rd ed. London: Taylor & Francis, 2000.

Trollope, Anthony. *Australia and New Zealand*, London 1873. Repr. Brisbane: University of Queensland Press, 1967.

Wallace, Alfred Russel. *The Malay Archipelago*, 1869. New York: Dover, 1962.

Weast, Robert C. (ed.). *CRC Handbook of Chemistry and Physics*, 58th ed., 1977.

Wells, H. G. *The New Machiavelli*, 1911. Project Gutenberg nmchv11.txt, accessed 3/11/04.

White, Mrs. Anna. *Youth's Educator for Home and Society*. Chicago: Union Publishing House, 1896, http://www.history.rochester.edu/ehp-book/yefhas/.

Wilde, Oscar. "Intentions," in *Pen, Pencil and Poison*, 16th ed. London: Methuen & Co., 1927.

Woffinden, Bob. "Cover-up," *The Guardian*, August 25, 2001, available at http://www.guardian.co.uk/weekend/story/0,3605,541260,00.html, accessed 3/13/04.

Xenophon. *Anabasis, c.* 360 BC. Project Gutenberg anbss10.txt, accessed 3/11/04.

Xenophon. *Cyropaedia: The Education of Cyrus*. Project Gutenberg cyrus10.txt, accessed 3/11/04.

Zimmet, Paul, Alberti, K .G. M. M., and Shaw, J. "Global and Societal Implications of the Diabetes Epidemic," *Nature* 414, 2001, pp. 782–87.

ACKNOWLEDGMENTS

Writers are the ultimate parasites, gathering in advice and information, absorbing support, and giving very little in return. Chris, Duncan (who dabbles in germs and things), and Threeby support me at home. Cate (who dabbles with heavy metals and pesticides and things) and Julian and Angus do so at a slightly greater distance. None of them poisons me, though they maintain the coffee supply. They encourage me, find me snippets of fine science, and put up with my temporary obsessions. I have a rare and marvelous family.

At Allen & Unwin, a host of Emmas give support, but especially Emma Cotter, aka Emma the Excellent Editor, while Ian Bowring thinks I do not notice the deft way he steers me away from my excesses. Both Emma and Ian have put themselves into this book as well, by the way they have helped me reconstruct. It takes a rare skill to conduct a prima donna. Emma also lent me books.

From one Internet list, Toby Fiander in Sydney instructed me patiently on the nature of tube wells, in order that I might understand better the arsenical tube wells of Bangladesh; Margaret Ruwoldt at the University of Melbourne introduced me to nineteenth-century medical jurisprudence and lent me books; David Allen found me new and exciting murderous herbs to romp through. Toby, David, Margaret, and Geoff Zero Sum saw an early draft and advised me well. Jean Lowerison, Theta Brentnall, Toby, David, and Margaret also saw a later version, patiently read it again and chided me, gently.

A number of these and other friends chipped in with comments about the last version: my son Angus, Theta, and Margaret (again!), Barbara Edlen, Barbara Sloan, Dee Churchill, Ian Musgrave, Mary Lou White, Paul Tessmer, and Kitty Park among them. I thank all of them, and anybody I have missed, because I would have missed them more if they had not commented!

So far as research was concerned, I relied mainly on the State Library of New South Wales and the main and branch libraries of my alma mater, the University of Sydney, and the resources of Project Gutenberg. As well, I had a bunch of friends and helpers on the Internet, and some remarkable Web sources to draw on. Assorted Banyanites helped: Mike Pingleton alerted me to TTX and zombies; Sylvia Milne in Chester and Mary Lou White in Washington, D.C., both found a way for me to get to the Old Bailey, as did Laurie Sarney. Robin Carroll-Mann found a difficult *Decameron* reference for me, and others on the Stumpers-L list chimed in as well, while Andy Meharg at the University of Aberdeen generously shared some lowdown on arsenic among the Fabians. I found him thanks in part to leads from Lena Ma at the University of Florida, who introduced me to the world of brake ferns as absorbers of environmental arsenic.

With fronds like these, who needs enemas?

INDEX